Applied Production Technology of Vegetables

— Author —

Pradeep Kumar Singh

Assistant Professor-cum-Junior Scientist
Division of Vegetable Science
Sher-e-Kashmir University of Agricultural Science and Technology
Kashmir – 190025 (J&K)

2016

Daya Publishing House®

A Division of

Astral International Pvt. Ltd.

New Delhi – 110 002

Cataloging in Publication Data--DK
Courtesy: D.K. Agencies (P) Ltd. <docinfo@dkagencies.com>

Singh, Pradeep Kumar (Assistant professor of vegetable science), author.
Applied production technology of vegetables / author, Pradeep Kumar Singh.
      pages cm
Includes bibliographical references and index.

ISBN 978-93-5130-942-0 (International Edition)

1. Vegetables India. 2. Vegetables--Technological innovations--India. 3. Horticultural crops--India.   I. Title.
SB320.8.I5S56 2016   DDC 635.0954   23

Published by   :   **Daya Publishing House®**
                         A Division of
                         Astral International Pvt. Ltd.
                         – ISO 9001:2008 Certified Company –
                         4760-61/23, Ansari Road, Darya Ganj
                         New Delhi-110 002
                         Ph. 011-43549197, 23278134
                         E-mail: info@astralint.com
                         Website: www.astralint.com

Laser Typesetting   :   **Classic Computer Services**, Delhi - 110 035

Printed at   :   **Thomson Press India Limited**

FOREWORD

The book **Applied ProductionTechnology of Vegetables** is the most modern and comprehensive manual on practical aspects of vegetable production and designed as per the latest syllabi of ICAR for the students, scientists and technocrats of the discipline of agriculture to persue their research, teaching and extension activities. So it is imperative to have elementary knowledge of basic principles and practices of vegetables for development of technologies having relevance to the farmer's problem.

The recent advances made from time to time, make it necessary that new book manual on vegetable production are written. I, particularly feel happy to welcome, the present book manual on vegetable production technology written by Dr. Pradeep Kumar Singh since this is perhaps first comprehensive book on vegetable production, which focused much attention on practical aspects particularly on nutritive values, types of vegetables based on botanical and other factors, nursery seed bed and inter cultural practices, vegetable seed and sowing methods, nursery management, micro-irrigation and fertigation in major vegetables, importance of mulching in vegetables, important varieties/hybrids in vegetable crops, seed production and testing procedure in vegetable crops.

Dr. Pradeep Kumar Singh efforts will benefit those who are working in vegetable production research. The excellent book manual he has written on vegetable production technology will be great useful in production of vegetable crops.

It therefore fulfils on important long felt need of educational institutes in agriculture and allied science subjects.

Dr. Tej Partap

Vice-Chancellor

Sher-e-Kashmir University of Agricultural Sciences and Technology-Kashmir 190 025 (J&K)

PREFACE

The basic knowledge of agriculture is essential, for non agricultural graduates, who peruse master and doctorate degree in different disciplines of agriculture. It is necessary for non-agricultural students to know about the vegetable crops and their behavior in laboratories and fields, to conduct research in the vegetable science based. There are good numbers of publication in vegetable production technology dealing with basic as well as applied aspects. Most of the publications, however, do not provide information for conducting the practical in remedial courses. Therefore, a need was felt to bring out a manual book on practical aspects of vegetable providing comprehensive information at one place. This manual book has been designed and planned based on the content of remedial courses to meet the requirements of the non agricultural graduates perusing master and doctorate degree on the universities. At the end of each chapter, exercises and assignment have been given with solved examples. The book is divided into 14 chapters viz. nutritive values, types of vegetables based on botanical and other factors, nursery seed bed and inter cultural practices, vegetable seed and sowing methods, nursery management, micro-irrigation and fertigation in major vegetables, importance of mulching in vegetables, important varieties/hybrids in vegetable crops, seed production and testing procedure in vegetable crops, post harvest losses in vegetable crops and procedure for varietal testing and releasing steps in India, Some informative appendix and lastly glossary which is useful for vegetable production technology etc has been given in most lucid and simple manner understandable to students while treating the subject, unnecessary details which sometimes freighter the students have been omitted but the interested students will find easy access to there once they go through the selected list of references

. It is my firm belief that this manual will be useful not only to the post graduate students majoring in vegetable science but also benefited to the Scientists and Teachers of different SAU's and Research Institutes of ICAR.

The authors are highly indebted to Dr. Tej Partap, Vice Chancellor, Sher-e-Kashmir University of Agricultural Science and Technology-Kashmir for his consistent encouragement, support and guidance to me for this and other assignments. I wish to thanks Dr. Sada Shanker Singh, Ex. Professor Department of Vegetable Science, NDUAT-Faizabad who have been more than generous in guiding and encouraging me for such activities.

I appreciate the efforts of publisher Astral International Pvt. Ltd of all the publication mentioned in the book and executing a good job in time. I am very much thankful to my family members who allowed me for writing work on holidays and late night hours for years together.

Pradeep Kumar Singh

Contents

Applied Production Technology of Vegetables

The Author

Dr. Pradeep Kumar Singh, M.Sc., Ph.D. and NET (ASRB) in Vegetable Science and presently working as Assistant Professor (Sr. Scale) in the Division of Vegetable Science, Sher-e-Kashmir University of Agricultural Science and Technology- Kashmir Shalimar Srinagar. Dr. Singh also served as Senior Research Fellow in Department of Vegetable Science, NDUAT Kumarganj Faizabad and Indian Institute of Vegetable Research Varanasi. He is specialized in the field of vegetable breeding, especially in Solanaceous (Tomato) and cucurbitaceous (Cucumber) vegetables. He has an excellent experience of 10 years in the field of Research and extension and 06 years experience in Teaching. As a Teacher in SKUAST-K he has taught more than 10 different courses to the student of Post graduate and doctorate classes, delivered several lectures in the training programmes on different topics related to vegetable sciences. As a prolific writer, he has published 06 book chapters 18 research paper, 70 abstracts, 15 review articles and 32 popular articles and 01 extension folders in books, journals and magazine of scientific repute and has written 02 book and guided students of M.Sc.

He has also worked as paper reviewers in several national and International Societies and also an editorial member in National and International Journals.

Dr. Singh is life member of several professional bodies and societies.

Chapter 1
Nutritive Value of Vegetables

Challenging Era of Vegetables

Present Scenario

Presently the two major concerns of developing countries are to overcome hunger and malnutrition. According to UN food organization, about 25 million people die from hunger and exhaustation in India,the ill effects of chronic malnutrition one of greater every year in India the ill effects of chronic malnutrition one of greater magnitude in low income groups fighting hunger and malnutrition has become global moral responsibility and a major concern in developing countries.it has been estimated that in the year 2020 the population of world will increase to 8 billion and 1.2 billion in India. Around 80 percent of the world population (6 billion) will be in developing world of which 2 billion people will be under nourished. The diversified and highly nutritive vegetables are of great importance in alleviating hunger and malnutrition the significance of vegetables in human diet and nutrition is well recognized and the number of vegetarians is increasing in the world due to greater awareness of good health and nutritious healthy food. At present India has a total population of one billion and the consumption of vegetables per caput per day is about 175g which is below the recommended nutritional requirement of 380g per caput per day. It is therefore necessary to increase the present vegetable production of 87.53million tones(1998-99)to at least 350 million to meet the dietary requirements of vegetables for the increased population of 1.3 billion in the year 2020.

Nutritional Requirement and Vegetable Consumption

The recommended dietry allowance (RDA) per capita per day by Indian Council of Medical Research (ICMR) for adult males includes cereals (475g), pulses(80g), green lady vegetables (125g), pulses (80g), green leafy vegetable (125g), other vegetables (75g), roots and tubers (100g) and fruits (30g). This amounts to 300 g of vegetables everyday to make our diet balanced. In rural house hold and those below poverty line, per capita vegetable consumption is very low, apprehended to be even lower than 40 g per day. Over whelming importance of vegetables in the balanced diet is marked by their enormous presence in the daily diet schedule. In some developed countries like Italy (593 g), Japan (523g), U.S.A (469g), Canada (428g) and Australia (346g).

Vegetables provide all nutrients components like carbohydrates, protein, fat, vitamins, minerals and water along with roughage which are the essential constituents of a balanced diet. Calorific value of vegetables is not much in comparison to the cereals and animal products. However, calorific requirement can well be supplemented by carbohydrates rich vegetables. The gloring example is potato staple food of Ireland. Other such vegetables like cassava, taro, yam, sweet potato etc. are considered as important carbohydrate food elements in the torrid regions of Africa and Latin America. Vegetables having higher vitamins and minerals content are called protective foods. Raw vegetables are particularly, rich in minerals and vitamins. Minerals complexes formed as a result of metabolism of the products of animal origin create an excess of acid inside the body which disrupts protective mechanism and metabolic processes, vegetables neutralize these substances and provide alkaline reaction for normal metabolism. Cellulose pectin and other constituents present in the fibers of vegetables help to clear the bowels, reduce constipation and promote digestion.

Computation of Calorie from Food

1g carbohydrate = 4 kcal

1g protein = 4 kcal

1g fat = 9k cal

Average Nutritive value of Vegetables

Considering the many components that go into human nutrition, it is difficult to assess the overall nutritive value of a crop. However, Grubbin (1968) proposed the average nutritive value (ANV) of vegetables based on the following empirical formula

G protein/5 + g fibre + mg Ca/100 + mgFe/2 + mg carotene + mg vitamin C/40

Adult Consumption Unit (ACU)

Energy requirement of a person depends on his mode of living. The energy is expressed as kilo calorie or calorie. Such daily requirement of calorie is termed as Adult consumption Unit or ACU. Calorie requirement of an adult individual of sedentary work is 2400 kcal and regarded as 1 ACU. The adult consumption unit varies with age, sex and work schedule of a person.

Computation of Calorie from Food

1 g carbohydrate = 4 kcal

1 g protein = 4 kcal

1 g fat = 9 kcal

Table 1.1: Average Nutritive Value (ANV) of different Vegetables
(After Grubben, 1978)

Name of the Vegetables	ANV	ANV per m^2
Tomato	2.39	101
Brinjal	2.14	51
Sweet pepper	6.61	173
Okra	3.21	43
Cucumber	1.69	68
Pumpkin	2.68	44
Watermelon	0.90	23
Amaranth	11.32	204
Water spinach	7.57	436
Chinese cabbage	6.99	180
Lettuce	5.35	79
White cabbage	3.52	120
Cassava leaves	16.67	870
Vegetable cowpea	3.74	23.0
Lima bean	4.88	25.0
Mung bean (Sprouted)	2.94	61.5
Hyancith bean (dry)	14.03	42.1
Onion	2.05	78.7
Carrot	6.48	107.6
Taro	2.38	40.0
Turnip	2.03	20.9

ANV per m^2 has been calculated based on average yield and edible portion per hectare.

So, 100 g of rice, containing 79.0 carbohydrates, 6.4 g protein and 0.4 g fat, supply 79.0 x4 + 6.4 x 4 + 0.4 x 9 = 345.2 kcal.

Similarly, 100 g potato, containing 22.6 g carbohydrate, 1.6 g protein and 0.1 g fat, supply 97.7 kcal of the required energy (kcal), 70 per cent must come from carbohydrate food, 10 per cent from protein and 20 per cent from fat. So, daily requirement of carbohydrate, protein and fat (main nutrient components) for an adult individual of sedentary work requiring 1 ACU, *i.e.* 2400 kcal will be:

Carbohydrate = 2400 x0.7/4 = 420 g

Protein = 2400 x 0.1/4 = 60 g

Fat = 2400 x 0.2/9 = 53 g

**Table 1.2: Daily Requirement of Main Nutrient Components
on the Basis of Calorie Need**

Group	Particulars	ACU	Calorie Requirement
Adult man	Sedentary work	1.0	2400
	Moderate work	1.2	2880
	Heavy work	1.6	3840
Adult women	Sedentary work	0.8	1920
	Moderate work	0.9	2160
	Heavy work	1.2	2880
	Pregnancy	1.5	3600
	Lactation	1.0	2400
Adolescent boys and girls	9-12 years	0.8	1920
(12-21 years) Infants	7-9 years	0.7	1680
	5-7 years	0.6	1440
	3-5 years	0.5	1200
	1-3 years	0.4	960

Source: ICMR.

If full carbohydrate is to be supplied by rice then the person has to consume 420 x100/79 = 531 rice (rice contain 79.0 per cent carbohydrate). Similarly, if the carbohydrate requirement is to be supplemented solely by potato then he has to consume 1858 g potato per day. Such exemplary supplementation by single food is unscientific and makes the diet unbalanced. So, the requirement of nutrient components is to be met through different foods *viz.*cereals, vegetables (300 g), vegetable oils, sugar or molasses, fruits and their optional foods like fish, meat, egg etc.

Calculation

Calculate the amount of potato to be consumed by sedentary adult man per day if the entire carbohydrates requirement is fulfilled by potato consumption.

Solution

Calories required by sedentary adult man = 2400 kcal

$$\text{Carbohydrate} = \frac{2400 \times 0.7}{4} \times (1\text{g carbohydrate} = 4\,\text{kcal})$$

$$\text{Carbohydrate} = \frac{100 \times 400}{22} \times (1\text{g carbohydrate} = 4\,\text{kcal})$$

100 g potato contains 22.6g carbohydrate contains 22.6 g carbohydrate

420 g carbohydrate will be provided by 1858 g of potato per day.

Chapter 2
Classification of Vegetable Crops

Vegetable crops consist of about 1200 species from 18 families.Among them, more than 860 crops can be classified according to following ways: Vegetable crops consist of 1200 species from 78 families. Among them more than 860 species under 59 families belong to dicotyledoneae and the remaining species under the rest 19 families to monocotyledoneae. In the tropical and sub tropical parts of the world about 90 species of vegetables crops are cultivated but hardly 15 are commercially important. Vegetable crops are not only botanically necessary to classify the vegetable crops for their recognisation. Vegetable crops can be classified according to the following ways.

(1) Botanical classification (2) Classification based on hardiness (3) Classification based on growing season (4) Classification based on the tolerance to soil reaction (5) Classification based on salt tolerance (6) Classification based on the plant parts used as vegetable (7) Classification based on methods of culture and (8) Classification based on respiratory rate of freshly harvested produce.

Species under 59 families belong to dicotyledonous and remaining species under the rest 19 families to monocotyledonous. Classification and grouping of these vegetables facilitates identification and understanding their common characteristics and growing conditions vegetables.

Botanical Classification

This is by far the best method of classification especially for vegetable scientists and breeders as it is based on botanical relationship among different vegetable crops. However from the standard point of vegetable cultivation, this classification would not be of much use because cultural requirements for the vegetable crops belonging to the same group vary widely except in few cases. All the vegetables crops come under the division Angiosperms of sub communities spermatophyte. However in the Northern part of India, immature leaves of some ferns (division pterophyte of sub community pteridophyte) are consumed as leafy greens. In the present classification, over 140 different types of vegetable crops under 5 monocotyledoneae and 20 dicotyledoneae families have been included (Table 2.1) in addition, green banana (*Musa paradisiacal* L., musaceae) green papaya (*Carrica papaya* L.caricaceae) and immature jack fruit (*Artocarpus heterophyllus* Lamk, moraceae) are popularly used as vegetables in Indian cuisine.

Classification Based on Hardiness

This classification is based on the endurance of the vegetables crops against frosts and accordingly the vegetable Crops may be classified as hardy, semi hardy and tender. Hardy vegetable crops can withstand frosts without injury whereas tender vegetable crops cannot and are even killed by light frost. The semi hardy vegetable crops are not generally injured by light frost.

Hardy Vegetable Crops

Asparagus, Broccoli, Brussels sprouts, Cabbage, Chive, Collards, Garlic, Kale, Knol-khol, Leek, Onion. Pea, Parsley, Radish, Rhubarb, Spinach, Turnip.

Semi Hardy Vegetable Crops

Beetroot, Carrot, Cauliflower, Celery, Lettuce, Globe artichoke, Palak, Parsnip, Potato.

Tender Vegetable Crops

Okra, Tomato, Brinjal, Chillli. All cucurbits, Cowpea, Cluster bean, Amaranthus, Taro, Sweet potato, Cassava, Yam, Drumstick, Elephant foot Yam.

Classification Based on the Growing Season

In the classification based on hardiness, tolerance to frost is the main criteria of classification. This classification cannot suggest the conditions under which the crops grow best for example some hardy crops like Asparagus, Pea, Turnip, Cabbage, etc. cannot grow under hot environment while other hardy crops like Radish, Onion etc. can tolerate hot environment much better similarly some tender vegetable crops like Brinjal, Muskmelon, Cowpea etc. cannot grow well in a cool weather even if there is no frost from these stand points classification based on growing season may be more useful for the farmers based on the distinct growing season. In India vegetables may be classified as summer or spring summer season rainy season and winter season vegetable crops. Summer and rainy season vegetable crops are tropical *i.e.* heat loving and heat resistant vegetable crops and winter season Vegetable crops are temperate *i.e.* cold resistant and/or frost resistant vegetable crops. In the plains of India summer or summer season prevails from February to June rainy season from June to September and winter season from October to January. However, in some areas these seasons are not so distinct.

Table 2.1: Vegetable Crops, their Botanical Name and Family, Edible Plant Parts and Chromosome Number

Name	Botanical name	Edible Plant Parts	Chromosome No. (2n)
Monocotyledonous			
Amarilidaceaa (Alliaceaa)			
Onion	*Allium cepa* L.	Bulb and sometimes leaves	16
Multiplier onion	*Alliumcepa* var. *aggregatum* L.	Small bulbs	16
Top onion	*Alliumcepa* var. *viviparum* (metz) Alef	Roots and above ground bulbils	16
Garlic	*Alliumsativum* L.	Bulb consisting of several bulblets called cloves	16-32 (2x -4x)
Leek	*Ampeloprasum* L. var *porrum* (L) (Syn. *Allium porrum* L.)	Blanched stem and leaves	32,48 (4x,6x)
Welsh Onion	*Allium fistulosum* L.	Slightly enlarged stem and leaves	16
Shallot	*Allium fascalonicum*	Young bulbs and green leaves	16
Chive.	*Allium schoenoprasum* L.	Slightly enlarged stem and leaves	16,24,32
Kurrat	*Allium kurrat*	Green leaves	32(4x)
Chinese chive	*Allium tubersome*	Green leaves	32(4x)
Araceae			
Taro (*Arum*)	*Colocasia esculenta* (L) Scott	Corm and cormel	28,42
Eddoe type	*C. esculenta* var antiquorum	(cormel bigger than mother corm)	(2x-3x)
Dasheen type	*C. esculenta* var. *globulifera* (syn. *C. esculenta* var. *esculenta*)	(cormel smaller than mother cormel)	42(3x)
Giant Taro	*Alocasia macrorrhiza* (L) Scott *indica* (Roxb) scott	Corm (large and cylindrical) Corm (large and cylindrical)	26,28
	Cucullata (lour) Scott	Corm (large and cylindrical)	
Swamp Taro	*Cyrtosperma chamisonis*	Corm (thin, cylindrical and thick)	26,28
Tannia	*Xanthosoma sagittifolium* Scott	Root like offsets Corm	26

Contd...

Table 2.1-Contd...

Name	Botanical name	Edible Plant Parts	Chromosome No. (2n)
Elephant Foot (*Suran*) Yam (suran)	*Amorphophallus campanulatus* Roxb Blume	Corm (big, round)	26,28
Chard	*Vulgaris* L. Var	Large leaves and fleshy leafstalk	18
Spinach	*Spinacea oleracea.*L.	Rosette leaves produced from Shorter stem	12
French spinach (orach)	*Artiplex hortensis* L.	Leaves and immature shoots	–
Pigweed (bathua	*Chenopodium album* L.	Leaves and tender twigs	36 (4x)
Compositae			
Lettuce	*Lactuca sativa* L.		
Head type	*L. sativa* var. *capitata* L.	Leafy heads	18
Leaf type	*L. sativa* var.	Non heading leaves	18
Loaf shaped head Cos type	*L. sativa* var. *longifolia* L.	Elongated leaves forming	18
Asparagus type	*L. sativa* var. *asparagina*	Young leaves and thick young stem	18
Chicory	*Cichorium intybus* L.	Leaves	18
Endive	*Cichorium endivia* L.	Leaves	18
Globe artichoke	*Cynara scolymus* L.	Soft, fleshy receptable of flower	34
Jerusalem artichoke	*Helianthus tuberoseus*	Heads or buds and thickened Bases of anvolucral bracts root tuber	102 (6x)
Terragon	*Artemisia dracunculus* L.	Leaves and early shoots	18
Salsify	*Tragapogon porrifolius* L	Roots	18
Black salsify	*Scorzonear hispanica* L.	Long, black roots and leaves	–
Spanish salsify	*Scolymus hispanica* L.	Long root	–
Dandelion	*Taraxacum officinale* Weber ex wiggers	Roots and leaves	–

Contd...

Table 2.1–*Contd...*

Name	Botanical name	Edible Plant Parts	Chromosome No. (2n)
Convolvulaceae			
Sweet potato	*Ipomeae batatos* (L)	Root tuber which is formed due to thickening of adventitious roots.	90(6x)
Water spinach	*Ipomea aquantica* Forsk (syn. *L. reptans* (L.) POIR-)	Young terminal shoots and leaves	–
Dioscoreaceae Yam (khamula)	*Dioscorea* spp.	Underground stem tuber	30-80 (3X-8X)
Greater yam	*D. alata* L.	Underground stem tuber	40 (4X)
Lesser yam	*D.esculenta* (lour) Burkill	Underground stem tuber	40(4X)
White Yam	*D. rotundata* (L.) Poir	Underground stem tuber	40 (4X)
Gramineae			
Sweet corm	*Zea mays* L. var *rugosa*	Soft immature kernel	20
Liliceae			
Asparagas	*Asparagas officinalis* L.	Young shoots up to 1-2 Inch below the soil which are called spear	20
Dicotyledoneae			
Aizoceae			
New Zealand Spinach	*Tetragonia tetragonioides* (Pall- O. kuntze (syn *T. expansa* murr)	Tender leaves and tops	
Amarantahceae			
Leafy amaranthus			
Amaranthus Badi chaulai	*A.tricolor* L.	Leaves, stems	32 or 34
Chotti chaulai	*A. blitum* L.	Leaves, stems	32 or 34
Nate sag	*A. vivdis* L.	Leaves, stems	32 or 34

Contd...

Table 2.1–Contd...

Name	Botanical name	Edible Plant Parts	Chromosome No. (2n)
Other types	*A. tritis* L.	Leaves, stems	32 or 34
	A. dubius L. etc.	Leaves, stems	32 or 34
Basellaceae			
Indian spinach	*Basella rubra* L. var *alba*	Fleshy stem and leaves	24
(Malabar night Shade)	*B. rubra* L. var *rubra* (red type)	Fleshy stem and leaves	24
Chenopodiaceae			
Beet root	*Basella rubra* L. var. *alba*	Thick, fleshy taproot (actually the upper lower portion of root hypocotyls and lower portion from taproot)	13
Palak (spinach beet)	*B. vulgaris* L. var. *bengalensis* Hort	Juicy leaves	18
Leaf mustard	*B. juncea* (L) Czern and Coss var *cuneifolia* Roxb	Tender leaf	36(4x)
Curled mustard	*B. juncea* var. *crispifolia*	Tender leaf	36(4x)
White mustard	*Sinapis alba* L.	Young shoots and leaves	24
Water cress (*Brahmi sag*)	*Nas turium officinale* R. Br. (Syn. *Rorippa nasturtium aquaticum*)	Tender mustard flavored topand leaves	32
Water cress (other type).	*N. microphylla* (Syn *R. microphylla*)	Tender mustard flavored top and leaves	64(4x)
Garden cress	*Lepidium sativum* L.	Leaves	16,32
Upland cress	*Barbarea vulgaris* R.Br.	Leaves	–
Sea kale	*Crambe maritime*	Blanches, tender leaves and shoots	–
Catran	*C. tatarica* busch	New leaves shoots and roots	–
Tuberous	*Sinapis juncea* var	Root tuber	–
Rooted Chinese Mustard	*Napiformis* pall. and Bios	Root tuber	–

Contd...

Table 2.1–Contd...

Name	Botanical name	Edible Plant Parts	Chromosome No. (2n)
Cucurbitaceae			
Cucumber	*Cucumis sativus* L.	Fruit (mostly placenta)	14
Muskmelon	*Cucumis melo* L	Fruit (mostly pericarp and little	24
snap melon(photo/phooti)	*C. melo* L. var momordica duth. and full.	Mesopcarp do (cracks and disgenerates at maturity)	24
Long melon (kakri)	*C. melo* L var utillissimus duth and full (Syn *C. melo* var. *flexuosus*)	(long,slender) used mostly in immature stage)	24
Netted melon	*C. melo* L. var *reticultalus*	(with netted rind)	24
Pickling melon	*C. melo* L. var *conomon* Mark	(small, eaten as apple including rind)	24
Ivy gourd	*Coccina grandis* (L) Voigt (Syn. *C. indica* Wright and Arn or *C. cordifolia*)	Fruit	24
Chow chow	*Sechium edule* (Jacq) (Chayote)	Fruit	28
Mitha karela	*Cyclanthera pedata* (L.) Schrad	Fruit	32
Cantaloupe	*C. melo* L. var	Fruit (with netted, watery or Scaly surface)	24
Cassava melon	*C. melo* L. var *inodorus*	Fruit (with netted, watery or Scaly surface)	24
Mango melon	*C. melo* L. var chito (Syn. *C. melo* L. var Dudaim naud	(Small, with musky adour)	
Small gourd (meki)	*D. melo var agrestis* Naud	(Small, with musky adour)	24
Gherkin	*Cucumis anguria* L.	(Small, with musky adour) (like small cucumber)	24
Water melon	*Citrulus lanatus* L. Matsum and nakai (Syn *C. vulgaris* schrad)	Fruit	24
Round melon	*E. lanatus* var *fistulosus* (Stocks) Mansf (syn *C. fistulosus* stocks or	Fruit	22
(Tinda)	*Praecitrullus fistulosus* (Pang)		
Pumpkin	*Cucurbita moschata* (Duch) Poir	Fruit (mostly pericarp and little Mesocarp)	40

Contd...

Table 2.1-Contd...

Name	Botanical name	Edible Plant Parts	Chromosome No. (2n)
Summer squash	*F. pepo* L.	Fruit (mostly pericarp and little Mesocarp)	40
Winter squash	*C. maxima* Duch	Fruit (mostly pericarp and little Mesocarp)	40
Buffalo gourd	*G. ficifolia bouche*	Fruit (mostly pericarp and little Mesocarp)	40
Bottle gourd	*Lagenaria siceraria* (Molina) Standl	Fruit	22
Bitter gourd (Balsam pear)	*Momordica charantia* L.	Fruit	22
Balsam apple	*M. balsamina* L	Fruit	28
Giant spine Gourd (kakrol)	*M. cochinchinensis* Spreng	Fruit	28
Ridge or ribbed gourd	*Luffa acutangula* (L) Roxb	Fruit	26
Sponge gourd	*L cylindrical* roem (Syn *L. aegypilaca* Mill)	Fruit	26
Pointed gourd (Parwal)	*Trichosanthhhhes dioca* Roxb	Fruit	22
Snake gourd (Chichinda)	*T. anguina* L.	Fruit	22
Wild snake gourd	*T. cucummerina*	Fruit	22
Wax gourd (ash gourd)	*Benincasa hispida* (Thunb) Congn	Fruit	24
Cruciferae (Brasskcaceae)			
Cabbage	*Brassssica oleraceae* L.	Head which is actually large	18
(white cabbage)	*var capitata f alba* DC	Compact leafy bud head (leaves are	18
Red cabbage	*B. oleracea* L. *var*	Red and having distinct wax coating head (leaves are red and having distinct wax coating)	18
Savoy cabbage	*B. oleracea* L.	Head (leaves are blisrered in appearance	18
Cauliflower	*B. oleracea* L. *var botrytis* L.	Curd which is the fleshy arrested inflorescence	18
Brussels Sprouts	*B. oleracea* L. *var gemifera* DC.	Small, immature heads (buds) which are borne in axils	18

Contd...

Table 2.1-Contd...

Name	Botanical name	Edible Plant Parts	Chromosome No. (2n)
Sprouting broccoli (broccoli)	B. oleracea L. var Italica Plenck	Loose curd-like structure consisting of green flower bud clusters and thick fleshy flower stalks	18
Kholkhol	B. oleracea var. gonylodes L. syn B. caulorapa or B. oleracea var caulorapa)	Above ground stem tuberwhich arises due to thickening of stem tissue above the cotyledone	18
Kale/collard	B. oleracea var acephala	Rosette of leaves and young shoots at the top of stem which resemble cabbage prior to heading	18
Chinese cabbage (pak-choi) (Syn B. chinesis)	B. campestris Subsp chinesis L. (Syn B. chinesis)	Long leafy elongated and compact head fleshy petiole and leaf	20
Chinese cabbage (pet-sail)	B campestris subsp. pekinensis (Iour) Rupr(syn B. pekinensis or B. chinese var pekinesis)	Loose leafy heads	20
Chinese kale	B. alboglabra bailey	Tender leaves and petioles	–
Mibumna Turnip	B. japonica, B.campestris ssp Rapifera metz (syn B. Rapa L. var glabra kitamura)	Rosette of eaves and petioles swollen root which is actually fleshy and thickened hypocotyls	20
Rutaaabaga (Swede)	B. napobrassica (L) Rchb (syn. B napus var. Napobrassica peterm	Enlarged and elongated tap root	38
Radish	Raphanus sativus L	Flesy swollen primary root and Hypocotyls, and leaf	18
Small radish	R. sativus L.Var	Flesy swollen primary root and Hypocotyls, and leaf	18
Chinese radish	R. sativus L. var Longipinnatus	Flesy swollen primary root and Hypocotyls, and leaf	18
Rat-tail radish	R. caudatus L. (syn R. sativus var. caudatus)	Pod (fruit)	18
Horse radish	Armoracia rusticana Gaertn mey	Fleshy, cylindrical roots and sometimes leaves	32(4X)

Contd...

Table 2.1-Contd...

Name	Botanical name	Edible Plant Parts	Chromosome No. (2n)
Euphorbiaceae			
Cassava (tapioca)	*Manihot esculenta* crantz	Tuberous roots	36
Chekurmanis	*Sauropus androgynus* Labitae	Green leaves consumed as pot herbs	–
Chinese potato (Country potato)	*Coleus parviflorus* benth (Syn. *C. tuberosus* benth)	Tuberous adventitious roots	–
Chinese artichoke	*Stachys siebeldi* Miq.	Long slender tuberous roots	–
Hoary basil	*Ocinum americanum* L. (Kalitulsi)	Leaves	–
Moringacea			
Drumstick	*Moringa oleifera* Lamk (Syn green pod sometimes *M. Pterygosperma* Gaertn)	Green pod and sometimes leaf	28
Polygonaceae			
Rhubarb	*Rheum rhaponticum* L.	Thick leaf stalk	44(4x)
Sorrel	*Rumex acetosa*	New leaves	–
Patience dock	*R. patientia* L	New leaves	–
Buck wheat	*Fagopyrum tataicim*	Tender tops	16
Portulacaceae			
Pusley(kulfa)	*Portulaca oleracea* L.	Leaf and tender stem	54
Ceylon spinach	*Talinum triangulare* wild	Leaf and tender stem	–
Rutaceae			
Curry leaf	*Murraya koenigii* (L)	Leaves for flavouring Different dishes.	18
Solananceae			
Potato	*Solanum tuberosum* L.	Stem tuber	48
Brinjal	*Solaum melongena*	Fruit (fleshy placenta)	24

Contd...

Table 2.1–Contd...

Name	Botanical name	Edible Plant Parts	Chromosome No. (2n)
Fruit round or egg shaped	*S. melongena* var *esculentum*	Fruit (fleshy placenta)	24
Fruit long and slender	*S. melongena* var *serpantinum*	Fruit (fleshy placenta)	24
Plant dwarf	*S. melongena* var	Fruit (fleshy placenta)	24
Tomato	*Lycopersicon esculentum* Mill.	Fruit (central fleshy placenta)	24
Common	*L.esculentum* var *communue* bailey	Fruit (central fleshy placenta)	24
Large leafed	*L. esculentum* var *grandifolium* bailey	Fruit (central fleshy placenta)	24
Upright	*L. esculentum* var *validum* bailey	Fruit (central fleshy placenta)	24
Cherry	*L. esculentum* var *cearsiforme* Bailey	Fruit (central fleshy placenta)	24
Pear	*L. esculentum* var *pyriforme* Bailey	Fruit (central fleshy placenta)	24
Leguminoceae			
Peas (garden)	*Pisum sativum* L. SSP *hortense*	Tender seeds	14
Peas field	*P. sativum* ssp *arvense*	Dry seed as pulse	14
Snowpea	*P. sativum* ssp *hortense*	Pods and seeds (edible podded)	14
French bean	*Phaseolus vulgaris* L.	Tender pod and dry seed (rajmah)	22
Lima bean	*P. lunatus* L.	Tender pods and seeds	22
Multiflora bean (runner bean)	*P. coccineus* L.	Tender pods and seeds	22
Tepary bean	*P. aculifolius* grey	Tender pods and seeds	22
Thicket bean	*P. polystachyus*	Tender pods and seeds	22
Adzuki bean	*P. angularis* (wild) Wright (Syn. *Vigna angularis* (wild) Ohwi and ahhashi	Tender pods and seeds	22
Sword bean	*Canavalia gladiate* (Gucav) DC.	Tender pods and seeds	22, 24
Gotani bean	*C. plagiosperma*	Tender pods and seeds	22

Contd...

Table 2.1–Contd...

Name	Botanical name	Edible Plant Parts	Chromosome No. (2n)
Jack bean	C. ensiformis (L.) dc	Tender pods and seeds	22
Valvet bean	Mucuna deeringiana (BOR) Merr (Syn. stizolobium) Deeringianum Dort)	Tender pods and seeds	22
Cluter bean	Cyamopsis tetragonolobus	Fruit (central fleshy placenta)	14
Winged bean	Psophocarpus tetragonolobus (L.) DC	Green pods seeds flowers and Even roots	18, 22
Broad bean (Faba bean)	Vicia faba L.	Green pods seeds	12
Hyacinth bean	Llablab perpureus (L.) Sweet (Syn. Dolichos lablab (L.) Roxb.	Green pods seeds	22, 24
Horse gram (Kulthi bean)	Dolichos uniflorus Kam(Syn dbiflorus Auct L.)	Dry seed	22, 24
Cow pea	Vigna unguiculata ssp	Tender pod immature	22
(Black eye Pea)	Unguiculata cv-gr.unguiculata (Syn V. unguiculata ssp. sesquipedalis)	Tender pod immature seed and mature seed	22
Asparagsa bean	V. unguiculata ssp unguiculata cv-gr sesquipedalis (Syn V. unguiculata ssp. sesquipedalis)	Tender pod	22
Soybean	Glycine max L.	Tender seed, dry seed	22
Yam bean	Pachyrrhizus erosus urbn (Sakula)	Root tuber	22
African yam Bean	Sphenostylis stenocarpa	Root tuber	22
African locust	Parkia clappertonia	Seed and fruit pulp	–
Bean			
Ground bean	Voandzela subterranean (L.)	Tender pod	–
Marama bean	Tylosema esculentum (Burchell) Schreiber (Syn. Bauhinia esculenta Burchell)	Root tuber	–
Agati	Sesbania grandiflora poir	Flowers and sometimes Leaves	24
Fenugreek	Trigonella foenumgraccum	Tender leaves	16

Contd...

Table 2.1–Contd...

Name	Botanical name	Edible Plant Parts	Chromosome No. (2n)
Malvaceae			
Okra	*Abelmoschus esculentus* (L.) Moench	Tender fruit	126,130,132 (Allopolyploid)
Martynaceae			
Martynia	*Proboscidea jussieul*	Green very hairy and fleshy pod	–

Summer or Spring Season Vegetable Crops

Brinjal muskmelon watermelon long melon snap melon round melon bottle gourd bitter gourd snake gourd ash gourd ridge gourd sponge gourd pumpkin summer squash winter squash cucumber okra tomato chilli cowpea cluster bean amaranthus Indian spinach pointed gourd water spinach etc.

Rainy Season Vegetable Crops

Okra, Brinjal, Chilli, Bottle gourd, Bitter gourd, Snake gourd, Ash gourd, Ridge gourd. Sponge gourd, Pointed gourd, Ivy gourd, Cowpea, Cluster bean, Hyacinth bean etc

Winter or Autumn-Winter Season Vegetable Crops

Cabbage, Cauliflower, Knol-khol, Brussels sprouts, Sprouting broccoli, Radish, Carrot, Turnip, Beet, Spinach.

Table 2.2

Name	Botanical Name	Edible Plant Parts	Chromosome No. (2n)
Currant Tomato (juslen) Mil.	L. pimpinellifolium	Fruit	24
Sweet Pepper	Capsicum annuum L	Fruit	24
Tabasco Pepper	C. frustescens L.	Fruit	24
Husk Tomato	Phesalis pubescens L.	Small round fruit which is enclosed Inside a thick husk	24
Tomatillo	P. mixocarpa Brot	Fruit	24
Tree tomato	Cyphomandra betacea	Fruit which tastes somewhat like tomato	24
Umbelliferae			
Carrot	Daucus carota L.	Enlarged and fleshy taproot	18
Celery	Apium graveolens L. var dulce (mill)	Leaf stalk and leaf for salad and soups	22
Celeriac	A. graveolens L. var. (Turnip rooted Celery)	Thick tuberous root for stews and soups	22
Leaf celery	A. graveolens L. var.	Leaves as a condiment garnish	22
Parsley	Petroselinum crispum (mill)	Leaves aromatic for flavouring	22

Contd...

Table 2.2–*Contd...*

Name	Botanical Name	Edible Plant Parts	Chromosome No. (2n)
Garnish and salad			
Turnip rooted	*P. crispum* var *tuberosum*	Swollen roots for stews and soups	22
Parsley	*Petraselinum crispum*	Swollen roots for stews and soups	22
Turnip rooted	*Chaerophyllum bulbosum* L.	Short swollen roots	–
Chervil	*Anthriscum cerefolium*	Short swollen roots	
Skirret	*Sium sisarum* L.	Bunch of roots that is produced from crown leaves	–
Chervil	*Anthriscus cerefolium*	Bunch of roots that is produced from crown leaves	
Dill	*Anethum graveolens* L.	Young leaves and stems as a garnish	22
Coriander	*Coriandrum sativum* L.	Young leaves as a garnish	22

Classification Based on Methods of Culture

This classification is the best for the growers because the vegetable crops requiring the same cultural practices (but not the same season for cultivation) are grouped together through they may be divergent botanically. However the crop species in the group like cole crops, bulb crops cucurbits peas and beans belong to the same family.

Group 1: Potato

Group 2: Solaneceous fruits
Tomato, Brinjal, Chilli, Sweet peeper.

Group 3: Cole crops
Cabbage, Cauliflower, Sprouting broccoli, Brussels sprouts, Knol khol, Chinese cabbage.

Group 4: Cucurbits
Pumpkin, Summer squash, Winter squash, Muskmelon, Watermelon, Cucumber, Bottle gourd, Ash gourd, Ridge gourd, Snake gourd, Sponge gourd, Bitter gourd.

Group 5: Root Crops
Radish, Carrot, Beet, Turnip, Horse radish, Rutabaga, Salsify, Turnip, Rooted chervil, Spanish Salsify, Black salsify.

Group 6: Salad Crops
Lettuce, Celery, Endive, Chicory, Parsley. Sorrel, Water cress, Garden cress Upland cress.

Group 8: Green and Pot Herbs
Pea, French bean, cowpea, Hyancinth bean, Cluster bean, Winged bean, Lima bean, Valvet bean.

Group 11: Sweet Potato

Group 12: Okra

Group 13: Pointed Gourd

Group 14: Chayote (chow chow)

Group 15: Tropical Perennial Vegetable
Drumstick, Agathi, Chekkurmanis, Curry leaf.

Group 16: Temperate Perennial Vegetable Crops
Asparagus, Rhubarb, Sea kale, Globe artichoke.

There is another classification based on the respiratory rate of the freshly harvested produce which is very helpful from the storage point of view of the vegetables. Palak, Onion, Garlic, French bean, Pea, Fenugreek, Lettuce, Potato etc. This classification of vegetable crops is not valid for the entire plains of India because of the variation in severity and duration of winter and the time an duration of rain for example Hyacinth bean is cultivated during September to January in the plains of West Bengal because of the prevailing mild winter here. Similarly Chilli, Brinjal, Tomato, Sweet Potato are cultivated extensively in the winter months in West Bengal. The growing seasons are being more flexible with the development of new adoptable varieties like high temperature tolerant varieties of Cauliflower, Cabbage and cold tolerant varieties of Tomato. In temperate or cool season vegetable crops different vegetable crops different vegetative parts (roots, stems leaves and buds) or immature flowers are edible plant parts excepting Sweet potato, Cassava, Yam, Taro and New Zealand spinach which are warm season crops. On the other hand, tender and ripe fruits are edible plant parts of warm season vegeatable crops French bean and Nroad bean are exceptions being cool season vegetable crops.

Classification Based on Tolerance to Soil Reaction

Vegetable crops can be classified into three groups according to their tolerance to soil acoidity as less tolerant, moderately tolerant and very tolerant (Table 2.3).

Table 2.3: Classification Based on Soil Reaction

Soil Reactions	Name of the Vegetable Crops
Slightly tolerant (pH 6.8–6.0)	Asparagus, Beetroot, Cabbage, Celery, Cauliflower, Okra, Spinach, Palak, Onion, Leek, Lettuce, Muskmelon
Moderate tolerant (pH 6.8–5.5)	Carrot, Bean, Pumpkin, Squash, Cucumber, Tomato, Brinjal, Garlic, Turnip, Parsley, Pea, Pepper
Highly tolerant (pH 6.8–5.0)	Potato, Chicory, Rhubarb, Sweet potato, Watermelon

Classification Based on Salt Tolerance

Vegetable crops may be categorized into three groups: sensitive, moderately resistant and resistant based on their tolerance to soil salinity (Table 2.4).

Table 2.4: Classification Based on Salt Tolerance

Salt Tolerance	Name of the Vegetable Crops
Sensitive (0.25)	Tomato, Snake gourd
Medium tolerant (0.50)	Okra, Chillies, Cabbage, Sweet potato
(0.75)	Cauliflower, Amaranth, Onion, Radish, Bottle gourd
Highly tolerant (1.00)	Ridge gourd, French bean
(1.25)	Bitter gourd, Ash gourd

Classification Based on the Plant Parts Used as Vegetables

In this approach vegetable crops are grouped to their edible plant parts.

Leaves

Cabbage, Lettuce, Spinach, Palak, Amaranthus, Fenugreek, Indian spinach etc.

Stem (Modified)

Knol khol, Asparagus, Cauliflower etc.

Flower

Sprouting Broccoli, Globe Artichoke Agathi.

Underground Plant Parts

Potato, Sweet Potato, Taro, Yam, Elephant foot, Beet, Onion, Garlic, Radish, Carrot, Cassava etc.

Fruits

Brinjal, Tomato, Chilli, Peas and Beans, All cucurbits, Okra etc.

This classification is not of much values because cultural operations for the vegetable crops in each group is not the same. For example cultural requirements for Potato Sweet potato and Onion are quite different and similarly that of Tomato, Okra and Cucurbits are not related. However cultural requirements of some vegetable crops which are grouped together are in some way similar as in case of Root crops.

Classification Based on Edible Parts of the Vegetables

Produce of commerce or part of the crop, which is consumed as vegetable (is called edible part) and form the basis of classification in vegetable crops as given in Table 2.5.

Table 2.5: Classification Based on Edible Parts of the Vegetables

Edible Part	Vegetables
Root and tuber	Radish, Carrot, Turnip, Beet root, Sweet potato, Horse radish, Parsnip, Cassava, Rutabaga, Salsify, Table beet, Potato
Stem	Onion, (bulb), Garlic (bulbils) Artichoke, Yam (corm), Turmeric (rhizome), Asparagus (spear)
Leaf	Amaranthus, New Zealand Spinach, Shallot, Chard, Cabbage, Chinese Cabbage, Chicory, Endive, Brussels sprouts, Lettuce, Leek Fenugreek, Spinach, Poi, Bathua, Coriander
Immature flower	Cauliflower, Sprouting broccoli, Globe artichoke
Immature fruit	Brinjal, Okra, Eggplant, Gherkin, Squashes, Sweet corn, Cucumber, Pea, Guar, Sem, Broad bean, Pointed gourd, Summer squash, Snake gourd, Bottle gourd
Mature fruit	Tomato, Pepper

Classification Based on the Major Climatic Regions

Based on the ability of the plants to flower, fruits and produce seeds in different regions of the world. The vegetables crops have been classified into two major groups (a) Temperate crops, (b) Subtropical and tropical crops. In the former case, the plant needs temperate climate or extreme winter to be able to flower and produce seeds. Here, the crops can be grown for vegetable both in the tropical and temperate regions but it would produce seeds wherever temperate climate prevails. For example, the cabbage is able to produce excellent head for vegetables in tropical plains but normally fails to produce seeds there. It produces seeds freely in the hills where temperate climate prevails. Similarly, cauliflower have been grouped now into four maturity groups. Varieties of first three maturity groups, known as Indian cauliflowers, It can temperate climate is required for seed production through, it is also cultivated successfully in tropical and subtropical regions during winter. In some crops like Carrot, Radish and Turnip there is a great variation among the varieties regarding their adaptability. The exotic or temperate varieties requires temperate climate to produce seeds, whereas the tropical or Asian varieties produce seeds freely in the plains. The crops have been classified on another important criterion. In regions where frost is a common feature during winter some of the crops which have grown otherwise fail to grow. This is utmost importance in the tropical plains. It is possible to grow Tomato, Brinjal, or Gourds during the winter in the plains of eastern and southern states but in northwestern states this is not possible because of the frost which kills the crop. Tropical crops flower fruit and produce seeds freely in the warm or tropical climate.

Practice to Learn: Classification of different vegetable crops

Objective:To adequate with the classification of different vegetable crops

Material: Plant samples

Classify the given Plant Samples

OBSERVATION SHEET

Sample No.	Name	Botanical Name	Classified Based on			
			Life Cycle	*Growing Season*	*Agronomic Use*	*Special Use Use*

Identification of Vegetable Crops

The vegetable plants differ each other in their morphological characters. The vegetative and reproductive parts of a plant help in clear identification. Some plants are very distinct whereas some can be distinguished on the basis of some very specific characters only. Keen and frequent observations make the identification easy. It is essential to know the different parts of the plants before undertaking the identification as these forms on the basis of distinguished characters. Some vegetables are very similar in their morphological characters and it is difficult to identify them especially during early stages of their growth.

Practice to learn: Identification of given plant samples on the basis of morphological characters and recording the observations in observation sheet.

Material: Plant samples of different vegetable crops.

Procedure:

1. Observe the plants critically.
2. Note different parts.
3. Sketch each plant showing portion of the stalk that bears leaf and label all the parts.
4. Draw the root system of the inflorescence of each plant and label the parts.
5. Record your observations in the observation sheet.

Practice to learn: Identifcation of morphologically similar plants and recording of their specific charcacters.

Record following observations in case of plants of Tomato, Brinjal, Chillies and Capscium.

OBSERVATION SHEETS

Characters	Tomato	Brinjal	Chilli	Capsicum
Plant growth habit				
Leaf type				
Fruit shape				
Fruit size				
Fruit colour				
Plant height				
Number of Primary branches				
Days to 50 per cent flowering				

Chapter 3
Nursery Seed Bed Preparation and Interculture

The nursery seed bed is the place where the seed germinate or it is a medium from which plant extract moisture and minerals through their roots. Therefore, the seed bed should have abundance of moisture, nutrients and air and to allow full penetration of plant roots. Regardless of the condition of the seed bed on the surface, it should be firm and reasonably compact beneath. A seed bed on the surface it should be firm and reasonably compact beneath. A seed bed properly prepared warms readily and holds a percentage of its water and plant nutrients in an available form. Moreover, the water and air movements are free and other things being equal to the plant grow well and give yields.

Tillage

Tillage involves the manipulation of the soil mechanically with the objective of promoting good tilth and in turn high crop production. Tilth refers to the physical condition of the soil in relation to the plant growth and hence must include all soil physical condition that influences crop development. Tillage is most important, difficult, labour and energy consuming operation in agriculture. Field preparation, which is the first operation of crop production, requires different types of tillage operations.

Objectives of Tillage

The objectives of tillage are as follows:

1. To break up the soil and to bring it into a suitable physical condition consisting of an aggregate of loose particles which will easily fall apart and crumble into the kind described as friable.

2. To remove roots, stubbles, weeds and other sprouting material like bulbs, stolons, roots etc to produce a clean soil medium.

3. To produce a condition favourable for ensuring adequate supply of moisture for the use of growing crops.

4. To destroy insect pests and worms and their eggs and larvae, which harbouring in the soil.

5. To leave the surface in such a condition to prevent erosion by wind.

6. Leveling of land for irrigation and other operations.

Types of Tillage

Primary Tillage

The operations performed to open up any cultivable land with a view to prepare seed bed for growing crops, is termed as primary tillage. Primary tillage is done mainly with heavy implements like plough, disc, harrows etc. Primary tillage usually serves as a basics for prepration of a good seed bed by subsequent operations like harrowing and leveling.The primary tillage equipments include mould board plough,disc and sub-soiler etc.

Secondary Tillage

Lighter and finer operations performed on the soil after primary tillage,but before and after seed placement are termed as secondary operations.These operations are generally done on the surface soil; very little inversion and shifting of the soil take place and consequently there is less power requirement per unit area. Secondary tillage implements include the harrows, cultivators, levelers and tillers. In many cases secondary tillage follows the deeper primary tillage operations.

The objective of secondary tillage are:

1. To improve the seed bed by greater pulverization of the soil.

2. To conserve moisture by summer fallow operations to kill weeds and reduce evaporation.

3. To cut up crop residue and cover crops and mix vegetative material with the top soil.

4. To break up clods, firm the top soil and put it in better tilth for seeding and germination of seeds and

5. To destroy weeds.

Tillage Implements

Mould Board Plough

The mould board plough is adapted to the breaking of many types of soil and is well suited for turning under and covering crop residues. The part of the plough that actually breaks the soil is called the bottom or base. It is composed of those parts necessary for rigid structure required to lift, turn and invert the soil. The parts which form the mould board plough bottom are the share, the land side and the mould board. The three parts are bolted to an irregular shaped piece of metal called the frog. The beam can also be attached to the frog.

Disk Plough

Disc ploughs are favoured in area where the climate is very dry and where the soil is rough and stony. Such soil conditions do not permit the operation of mould board ploughs to good advantage. They work well in variety of soil conditions ranging from heavy clay hardpan to loose sandy soils. It leaves the trash on top of the ground so as to conserve moisture. It is also preferred for land infested with heavy growth of vegetation and for land requiring deep ploughing for vegetation and for land requiring deep ploughing.

Chisel Plough

The chisel type plough is a tool with a rigid carved or straight shank with relatively narrow shovel points. It may be termed heavy duty deep cultivators. The furrow opened by the plough may be as close as 30 cm or as wide as 60 to 90 cm. The depth of ploughing may be as close as 30 cm or as wide as 60 to 90 cm. The depth of ploughing may be as shallow as desired as deep as 40 cm or more. The soil is broken by stirring and is not inverted and pulverized to the extent as mould board and disk ploughs crush the soil. This type of the plough is used for stubble much a sub surface tillage practice. It is also useful in breaking the hard layer of the soil just below the regular plough depth.

Sub Soilers

Sub soil ploughs are built heavier than the chisel ploughs. Since they are used to penetrate the soil to a depth from 50-100 cm, tractor of 60-85 HP may be required to pull a single standard ripping through a hard soil at a depth of 90 cm; large hydraulic cylinders are provided for lifting the ploughs Sub soil ploughs are available in both training and mounted units.

Desi Plough

Desi ploughs are animals drawn. They tear the soil surface but do not turn the soil. They form V shape furrows and may leave some un-ploughed area between two furrows.

Puddlers

Puddling of soil is one of the most common farm operations in paddy growing areas. This usually refer to the churning of soil in the presence of water by means of a puddler for the purpose. The main purpose of puddling is to reduce leaching of water, to kill weeds and make the soil soft enough for the transplanting of paddy seedlings. The puddlers may be animals or tractor drawn. Following paddlers are used.

Desi Plough

It is the most prevalent implement used as a puddler in paddy growing areas, although it does not do the job effectively.

Open Blade Type Puddler

This is commonly used for puddling in south India. It consists of series of steel or cast iron blades fastened to a cast iron hub. The cast iron hubs are two or three in number. These hubs revolve on steel shaft to which wooden beam or operator seat is attached. The action of this implement is much better than the other animal drawn implements for the purpose. Tractors mounted with cage wheel and attached with harrow or cultivator or rotavator perform the job of pudding satisfactory.

Power Rotary Tiller

This is mounted on the rear side of power tiller. The blades mounted on the rotary shaft cut soil and prepare the field for transplanting of paddy.

Harrows

Harrows are those tillage tools which are used to prepare the land by breaking clods, cutting weeds, pulverizing the soil, covering seeds and smoothening the surface. They are often used in seed bed preparation just before planting. There are different types of harrows.

Disc Harrows

The disc harrow does an effective job of cutting and covering. The function is to pulverize the soil leaving the surface soil mulch and a compact sub-surface. The harrow has a concave blades which can be made to cut either shallow or deep by mechanical or hydraulic controlled mechanism. The number of discs in a gang varies from 4-10 spaced at 15-25 cm apart and their diameter ranges between 40-60 cm.

Drag Harrows

Drag Harrows are being used since ancient times, early farmers used to cut branches from the trees for use in leveling the fields. Even today in some places farmers drag bamboo pieces with long nails to break the soil crust and stir the surface. These harrows are used to break the clods, to stir the soil, to uproot the early weeds, to level the ground, to break the soil crust and to cover the seed and similar other operations.

Spike Tooth Harrow

It is pointed steel pegs about 2-3 cm long with their pointed end towards ground. The each peg is rigidly clamped with the help of U-bolt to the cross bars frame.

Spring Type Harrow

Tractor drawn harrow have looping, elliptical or spring like types. They are used extensively to prepare ploughed land before planting.

Blade Harrow

This harrow is popularly known as *bakhar*.It is the most commonly used by Indian farmers and generally used in clay soils for preparing seed beds in both *kharif and Rabi* crops. The action of blade harrow is like that of sweep, moving into the top surface of the soil without inverting it. Sometimes, it is used to chisel out the uncut portion left after ploughing by an indigenous plough.

Cultivators

Cultivators usually refer to the tillage operation of manipulating the soil after the seed is planted or the seedlings have emerged. The cultivators are used for the following purposes:

☆ To control weeds so that they do not compete with crop plant growth factors.

☆ To prevent/reduce surface evaporation.

☆ To maintain the seed bed in good tilth during the growth of the crop.

☆ To achieve rapid infiltration of rainfall and adequate aeration.

The commonly used cultivators are either animal drawn or tractor drawn. The tractor drawn cultivators are mostly mounted, generally front mounted types. They can be quite accurately adjusted before being used in the field and can be easily used to cultivate within 7.5 cm spacing between the rows. The width covered by the cultivators depends upon the row spacing of crop.

Clod Crushers and Levelers

These implements are used immediately after ploughing or harrowing if the land is to be prepared for seeding. The main purpose of operating these implements is to crush, grind and tear the unevenly ploughed soil to produce smooth, well packed seed bed. It also reduces evaporation losses from the land surface.

The most common clod crusher is Pata or Patela (Planker) which is generally a rectangular section of long wooden log provided with two pegs for hitching. The length of pata depends upon the power of the bullocks.

In field preparation, leveling is an essential operation. Leveled fields receive uniform penetration of irrigation water with high efficiency. The possibility of water logging and soil erosion are reduced considerably. The entire field becomes ready to receive timely agricultural operations like ploughing, seeding and inter-culture. Smooth fields also facilitate the operations of field equipment and are very important for mechanical harvesting. The land leveling is done with wooden logs or planks by the farmers. They are operated in ploughed land to collect loose soil from high spots and to dump it into depressions. The other type of leveler is called Karha or Scraper, which is slightly improved over the wooden planker. It consists of depressed steel sheet made of 3mm thick mild sheet. Its cutting edge is generally made of high carbon steel. It is also provided with a wooden handle in the centre. The implement is pulled by a pair of bullocks. The amount of work done in a day depends upon the various factors such as hardness of the soil, transportation distance and the volume of the soil cut each time.

Special tractor drawn land planes consists of a long frame, supported by two wheels, and an adjustable leveling blade, fitted at the intermediate point. They are provided with steel wheels so that the implement may not penetrate much into loose soils. They are commonly used for leveling irrigated fields.

Hand Tools and other Implements

City farming needs a few hand tools and implements for carrying out day-to-day garden operations like digging, leveling, handling, cutting, spraying, pruning, mixing and transporting manure and water, etc. A number of multipurpose garden tools are available in the market. Multipurpose tools are preferred to single purpose tools for garden operations. The commonly used hand tools and implements and their applications are given below.

Pick Axe, Spade and Crowbar

☆ **Pick Axe**: It is a relatively heavy tool used to dig trenches, make water channels and put earthen support to walls. It has various shapes, sizes and weights with a long wooden handle.

☆ **Spade:** It is a hand operated tool designed for digging or removing the earth. It has a long handle and a thick flat blade that can be pressed into the ground with the help of the foot.

☆ **Crowbar:** It is an approximately 120 to 150 centimetres (cm) long steel bar usually approximately 3 to 5 cm in diameter and sharp at one end and blunt at other. It is used to dig holes in the ground to fix poles for pulling overhead trellis.

Hand Cultivator, Hand Fork and Hand Hoe

☆ **Hand Cultivator:** It has three to five round shaped blades and is used for stirring and pulverizing soil before planting seeds or transplanting seedling in a Nursery seed bed. It can also be used to remove weeds from the seed bed and aerate and loosen the soil in the standing crop.

☆ **Hand Fork:** It is made up of steel and normally has four prongs with a wooden handle and is used in pulverizing and stirring subsoil in the kitchen garden beds.

☆ **Hand Hoe:** It is made of a sharp metal blade attached to a long wooden handle, used for removing weeds and unwanted plants. Along with the spade and fork the hoe is also considered as a basic garden tool.

Garden Rake, Secateur and Hand Saw

☆ **Garden Rake:** It is a garden hand tool made up of a toothed bar fixed with a metal pipe to a long wooden handle and used to collect leaves, grasses and hay. It is most useful for loosening soil, light weeding and leveling.

☆ **Secateur:** It can also be called a hand pruner. It is very hard and can prune hard branches of trees and shrubs up to 2 cm thickness. A secateur has two blades and a short handle and is operated by one hand. The powerful spring between the two handles causes the jaw to open again and again after pruning. When not in use it can be locked with a safety catch.

☆ **Hand Saw:** It is one of the very useful garden tools used for pruning unwanted branches of growing trees in the garden. It has a serrated broad blade usually ranging from 30 to 60 cm long with a handle at one side.

Hand Spray Pump, Watering Can and Wheelbarrow

☆ **Hand Spray Pump:** It is a round bottomed plastic pot fitted with a nozzle at the top of the pot and a handle at the side of the body. It usually accommodates 1 to 2 liters of water used for spraying plain water or pesticides on garden vegetables.

☆ **Watering Can:** It is a metal pot attached with a handle at one side and a long pipe with a broad sieve at the other end and used to water or irrigate crops in the kitchen garden. Its water carrying capacity varies from 5 -10 liters.

☆ **Wheelbarrow:** It is a rectangular vehicle with one wheel or two normally pushed forward. It is used to transport garden soil, manures, earthen pots and other heavy materials in the kitchen garden.

(a) Practice to Learn

Prepare the seed bed provided and record your observations

OBSERVATION SHEET

Item	Observations
Implement used	
Operated by	
Area prepared	
Time taken for preparation	
Moisture condition	
Weeds situation	
Previous Vegetable crops	
Describe condition after preparation field	
Any other	

(b) Practice to Learn

Identification of agricultural implements and their categorization examine the implements shown and record the observations in the given observation sheet

OBSERVATION SHEET

Sl.No.	Name of the Implement	Operations for which Used	Driven by	Specific Characters for Identification

Chapter 4
Vegetables Seed and Sowing

Seed

Seed can be considered as the starting organ in the life of higher plants. It germinates to give rise to the seedling and eventually matures in to a plant. The seed is viable even if kept as such for months or even years while other organs of plants have relatively shorter longevity. Thus, information of seed prolongs life of a plant.

In general, the seeds are either monocotyledonous or dicotyledonous. In the vegetable crops belonging to the Lilium family (Amarllidaceae), the seed unit is actually the fruit. The Seeds of this family are monocotyledonous with endosperm. The crop seeds of *Leguminosae* family are true seeds. The seeds of this family are dicotyledonous and are without an endosperm.

Identification of Vegetable Seeds

For successful vegetable production, it is essential to maintain the quality of seeds and plants. Significance of crop seed identification cannot be over-emphasized. The seeds can be identified on the basis of their morphological characters like shape, size, colour, shine etc. The main objectives of this to familiarize the students with the prominent characteristics of seeds useful in their identification. Some of the morphological characters useful in identifying seeds are:

Vegetable Seed Coat

The testa or outer layer of the seeds known as seed coat. It may be hard, thick, thin, papery or brittle and is developed from the integuments of the ovule.

Vegetable Seed Size

Seeds are identified as small, medium or large (big or bold) size. This is relative term. Small size seeds are that of cabbage, cauliflower, kale, tomato, chilli, onion. Large size seeds are of bitter gourd, beans, peas etc.

Vegetable Seed Shape

The shape of different vegetable seeds varies to a great extent. The shape may be spherical, oval, elliptical, flat winged, elongated disc like, kidney shaped etc. This is very important characters in identification.

Vegetable Seed Surface

The surface of the seed may vary from smooth and glossy to dull or rough. There may be some wrinkles, reticulate markings, coloured spots etc. The surface may also be covered with some out growth in the form of spines, hooks etc. Presence of such marks on the surface makes the identification easy.

Vegetable Seed Colour

All parts of the spectrum may be represented in the colour of the seeds. The seed may be uniform colour or may be mottled, streaked or spotted. The colour may be dark brown, reddish brown, yellowish brown, purple, light brown, cream, amber, brightly coloured or dull.

Hilum

It is scar-like structure where the seed breaks off from the funicle. Funicle is a short stalk which attaches the seed to the pod. In leguminous vegetables it may be small, circular, oblong, cleft or wedge shaped. It is prominently visible in some seeds like peas, beans etc.

Micropile

If a soaked seed is pressed, a drop of water will ooze out from a small pore below hilum. This pore is termed as micropile. In some seeds, a songy structure turned crauncle the micropile less distinct.

Raphe

Directly above the hilum, there is raised ridge visible on the seed called raphe. The raphe may be very prominent in some seeds whereas less prominent in others.

Chalaza

A little away from the hilum, a dark triangular patch on the raphe is called chalaza. Its location on the seed is a very prominent character for identification.

Practical Activities

On the basis of morphological characters of the seed, identify the given samples and record your observations in the given table 4.1.

Material required: (a) Vegetable seed sample (b) Lens

OBSERVATION SHEET

Name of the Vegetables	Botanical Name	Monocot/ Dicot	Size	Shape	Surface	Colour	Luster	Any other Specific Characters
Tomato								
Brinjal								
Chilli								
Capsicum								
Caulifloower								
Cabbage								
Knol-khol								
Broccoli								
Radish								
Carrot								
Turnip								
Beet root								
Cucumber								
Bottle gourd								
Bitter gourd								
Water melon								
Musk melon								
Pointed gourd								
Snake gourd								
Peas								
French bean								
Broad bean								
Winged bean								
Dolichos bean								
Cow pea								
Lettuce								
Onion								
Garlic								
Okra								
Spinach								
Fenugreek								
Coriander								
Amaranthus								
Potato								

Methodology

☆ Spread the seeds of a crop on a glass plate and observe them with the lens and note various identification marks particularly the location of the hilum, raphe, chalaza and the tuff of hair at the apex, wrinkles, dots, spots etc on the surface.

☆ Make sketches of all types of seeds

☆ Record your observations in the observation sheet.

Sowing

Sowing is the most important operation in successful vegetable production. Time and methods of sowing play crucial role in successful crop raising. The time of sowing of different vegetables varies according to the requirement of the vegetable crops for temperature, radiation and other growth factors. The method of sowing can be modified and adopted to achieve the optimum productivity of different vegetable crops.

Methods of Sowing

The common methods used for sowing of different vegetable crops are (a) Dibbling method, (b) Seed dropping behind the plough, (c) Drilling, (d) Hill dropping, (e) Broadcast sowing, (f) Check sowing and (h) Transplanting.

Dibbling Method

This means placing two or more seeds in holes made in the soil either by hand tools or by some implements. Dibbling of seed is only done for small pots and is generally used for vegetable crops. The method is time and labour consuming but seed requirement is reduced considerably. Good for high value vegetable crops where seed available is less and cost is high.

Seed Dropping behind the Plough

Seeding behind the plough in the furrows is used for larger seeds like peas, beans. The usual method is to use indigenous plough to open a furrow in which a man women following the plough drops the seeds by hand. When the next furrow is opened the previous furrow gets partially covered. The limitation of the methods is that distribution of seed in each furrow may not be uniform as it depends on the skill of the person dropping the seed. The depth of furrow and distance between furrows may also vary.

Drilling

Dropping the seeds in the furrow through seed tubes is termed as drilling. Calibration of seed may either be done manually or mechanically. Some of the mechanically operated seed drills give a very high accuracy. The number of rows planted at a time may be one or more. In this method, accuracy of proper depth, spacing and amount of seed sown is much higher than the other methods. The area covered per day is also more. The only disadvantage is that mechanical seed drills are likely to get clogged during operation and this may result in irregular germination of the vegetable crops.

Hill Dropping

The seeds are sown in lines as in drilling and the seed dropping in line is also controlled. Unlike drilling, the seeds are dropped at a fixed spacing, not in a continuous stream. Thus; the spacing between the plants in a row is constant. But the spacing between rows is not the same as that between the plant in a row.

Broadcast Sowing

Broadcasting is the scattering of seeds on the field surface. Soon after broadcasting the seeds are covered by manipulating the soil and planking it over. Such vegetable crops are given interculture operations only by hand tools, like Kurpi, Kudali etc. A slightly higher seed rate is required in this method as seed remains on surface or placed too deep suffer adversely.

Transplanting

Transplanting of seedlings is commonly done for vegetable crops like tomato, brinjal, chilli, capsicum, lettuce, onion etc. In general transplanted vegetables gives higher yield than that planted otherwise. Perhaps it is due to better care given to seedlings in the nursery and plants in the main field. Transplanting is done by manual labour in India. It is the most time consuming operation which is not much liked by the labourers. Transplanting may be done in lines or otherwise. However, it is done in line then the interculture operations are facilitated.

Seeding Implements

Seed Drill

A seed drill performs the following mechanical functions:

☆ It opens furrows at a uniform depth.

☆ It drops seeds uniformly without injury.

☆ It covers the seeds and compacts the soil around them.

☆ In addition, the seed drill equipped with fertilizer attachment distributes the fertilizer evenly by the side of row in which seeds are placed.

Depending upon their make, seed drill may be classified as:

Manually Metered Seed Drill

It is simplest seed drill which the Indian farmers are using, which consists of: A seed bowl or funnel and vertical seed bowl fitted to indigenous (desi) plough.

The vertical tube is either fitted to the shoe of the plough or it is tied with the body to drop seeds just behind the plough in the furrow. When the tube is connected to the shoe, a hole is drilled into the shoe for seeds to drop in the furrow. There are seed drills in common use equipped with two, three, four, six, eight or even twelve seed tubes. They are drawn by a pair of bullocks. In manually metered seed drills, the uniformity of the seed distribution depends upon the skill of the man dropping the seed in the bowl. There is no seed-metering device to control the desired amount of seed to be shown.

Mechanically Metered Seed Drill

Mechanically metered animal drawn seed drills are available with a maximum of about six furrow openers. The main parts of such a seed drill are:

- ☆ Seed box
- ☆ Metering mechanism
- ☆ Seed tubes
- ☆ Furrow opener
- ☆ Raising and lowering devices
- ☆ Seed adjustment levers
- ☆ Transport-cum-power transmitting wheels

Tractor Drawn Seed Drill

They are trailed type, semi mounted type or mounted type. Depeding upon the type/size of tractor, these machines are available in different models. The size is usually expressed by the number of furrow openers and their spacing. The spacing of furrow openers on a machine is fixed at about 19 cm. Sometimes alternate seed tubes are closed to increase the spacing between rows.

Seed-cum-Fertilizer Drill

The seed box of this device is divided into two compartments, the bigger one for the seeds and the smaller for the fertilizer. Most fertilizer drills have provision to drop fertilizer either though the same tube along with seed or through a separate tube directly behind the seed tube.

Planters

Planters are used for row drilling of larger seeds than those, which normally go through seed drill. They give more accurate results with larger seed. Rows are far apart to allow inter-cultivation. Like the seed drills, a furrow opener is provided on the planters. The furrow openars are called runners. On most planters, the peak, wheel is provided both for covering the seed and making the soil firm over the planted seed. Some of the tractor drawn peas planters use a plough bottom to open furrow and two small shovels to cover the seeds. This type of arrangement is common in dry areas where moisture conservation is the main aim. Planters are provided with a seed hopper for each row. There may be separate hopper for fertilizers.

Prerequisites for Good Crop Establishment

Proper Depth of Seed Placement

Vegetable seeds must be planted at the proper depth if adequate stand is to be aehieved. Seed size is generally, related to the optimum depth. Seedling from large seeds will usually emerge from greater depths than seedlings from small seeds for the following reasons:

- ☆ Large seeds have a larger food reserve which can be utilized in the emergence process.

☆ The elongation potential of seedling from large seed is usually greater *i.e.* the hypocotyl (epigeal emergence) or mesocotyl (hypogeal emergence) has a greater possibility for elongation.

The type of emergence exhibited by a species also influences the optimum depth of planting. Crop of a given seed size with epigeal emergence should generally not be planted as deeply as crop with hypogeal emergence because of the difficulty of pushing the cotyledons through the soil. The depth of planting is also influenced by soil type. Deeper plantings are usually possible in lighter soils and in soil with better structure than in heavier soil types.

Good Seed and Soil Contact

Good seed and soil contact depends upon methods of sowing, implement used and tillage operations prior to sowing.

☆ Tillage prior to planting should ensure a seed bed free of clods and firm enough so that the seed regardless of size can have a large part of its surface in contact with soil.

☆ The planting equipment used should provide for good seed soil contact without packing the surface layer of the soil. Such packing of the surface layer often results in formation of "crusts" which may inhibit emergence.

Optimum Plant Stand

The arrangement of plant is very important. Some of the factors which affect planting rate and distribution are:

☆ Percent germination and purity of seed lot and number of viable seeds of the crop are very important in determining the desired stand.

☆ Competitive ability of the plant is important to utilize space. The larger plants should usually be planted at lower population with wider space distribution.

☆ Species or the varieties which have a greater tendency to branches are planted at lower population. Branching provide a method for the plant to take advantage of the surrounding space.

☆ Competitive ability of plant or varieties mixed cropping determines the seed rate.

Time of Planting

In general, any delay in planting after an optimum data results in decreased crop yield. The crop plant needs time to reach at optimum vegetative growth stage for sunlight, interception, photosynthesis and to produce dry matter. Therefore, planting at optimum time usually give higher yield when other factors are equal.

Practical Activity

OBSERVATION SHEET

Items	Vegetables			
	Tomato	Brinjal	Cabbage	Bottle Gourd
Row spacing (cm)				
Depth of sowing (cm)				
Seed rate (g/ha)				
Methods of sowing				
Implements used				
Methods of fertilizers				
Precaution taken				
Any other details				

Seed Requirements

In vegetable production use of quality seed is essential input for high productivity. Good quality seed should have following characteristics:

Genetic Purity

The seed should be genetically pure and true to type. To ensure this, the seed should always be procured from reliable source.

Physical Purity

The seed should be free from mixture of other crops, weeds, chaffs, stones, soil particles, straw etc. Mixture with weed seeds lead to several problems. These extra materials affect the seed quality influencing the seed requirement and germination percentage.

Uniformity in Seed Shape, Size and Colour

The seeds should be uniform in their size, shape and colour. Seeds of different sizes affect productivity. Smaller seeds are either not fully mature or diseased. Small and shriveled seeds give rise week seedlings.

Seed Maturity

Seeds should be obtained from fully matured vegetable crops. Seeds obtained from pre-matured vegetable crops are poor in quality.

Seed Boldness

In this case, many aspects are taken into consideration. The seeds should not be damaged by insects during threshing. Dull spots on the surface indicate that the seed is not free from disease or insect damage. Ruptured or broken seeds are easily attacked by soil born disease.

Age of Seed

Seeds should be new and not very old. The old seeds lose their viability and germinability. Old seeds comparatively less shining then new seeds and hairy like structure less appear in this type of seeds.

Germination Capacity

The germination capacity of any seed is the most important quality character. The seed rate is decided this property of the seed. The seeds with higher germination percentage are better for sowing.

Seed Viability

If embryo is damaged, then the viability of seed is lost. Sometimes due to insect damage or extra moisture with high temperature during storage reduced seed viability. It is essential to know before sowing whether seed is viable or not.

Practical Assignment

Find out the physical purity of the given sample.

Materials

1. Seed samples
2. Petri dish
3. Weighing balance

Procedure

Weigh the sample given on the weighing balance.

Separate out the good seeds, straw, dust, stones etc.

Weigh the different components obtained

Record these observations in following observation sheet

OBSERVATION SHEET

Observations	Vegetables Seed Samples		
	1	2	3
Total sample weight (g)			
Weight of seed only (g)			
Weight of weed seeds (g)			
Weight of chaff (g)			
Weight of other materials (g)			

Work out physical purity (PP) as under

$$PP \text{ (per cent)} = \frac{\text{Weight of pure seeds}}{\text{Total weight of sample}} \times 100$$

Or

$$PP \text{ (per cent)} = \frac{\text{Total weight of sample} - [\text{Weight of weed seeds} + \text{weight of chaff} + \text{weight of other material}]}{\text{Total weight of sample}} \times 100$$

Practice to Learn

Find out the germination percentage of given seed sample.

Materials

1. Seed sample
2. Petri dishes
3. Towel papers (filter paper)
4. Forceps
5. Seed germinator

Procedure

The germination test can be held by two methods:

(a) Petri Dish Method

Take a clean dish, keep a filter paper at the bottom and soak it with water. Spread the 10 or 20 seeds depending on size on moist filter paper. Cover the petri dish and keep in seed germinator. Add little water on second day and subsequently, if required. After a week observe the seeds and record your observations in given observation sheet.

OBSERVATION SHEET

Observations	Samples No.		
	1	2	3
No. of seeds kept for germination			
No. of seeds germinated (normal) with properly developed radical and plumule			
No. of seeds not germinated (hard seeds)			
No. of seeds with normal radical and abnormal plumule			
No. of seeds with abnormal radical and normal plumule			
No. of seeds with abnormal radical and abnormal plumule			

$$\text{Germination (per cent)} = \frac{\text{No. of seeds with normal germination}}{\text{No. of seeds kept for germination}} \times 100$$

Or

$$\text{Germination (per cent)} = \frac{\begin{array}{c}\text{Total (Total No of seeds kept for germination)} - \\ \text{(No. of hard seeds + No. of seeds showing} \\ \text{abnormal germination)}\end{array}}{\text{No. of seeds kept for germination}} \times 100$$

(b) Towel Paper Method

Take pieces of towel paper and soak them in water. Spread the towel paper on table. Arrange the seeds in rows and columns of 10 each with the help of a forceps. Cover the seeds with other wet towel paper. Roll the sheets and fix with a rubber band and keep in germination. After a week open the rolled papers count the seeds and work out the germination percentage as explained earlier.

OBSERVATION SHEET

Observations	Samples No.		
	1	2	3
No. of seeds kept for germination			
No. of seeds germinated (normal) with properly developed radical and plumule			
No. of seeds not germinated (hard seeds)			
No. of seeds with normal radical and abnormal plumule			
No. of seeds with abnormal radical and normal plumule			
No. of seeds with abnormal radical and abnormal plumule			

Practice to Learn

Carry out the viability test and work out the percent viability of different seed samples.

Materials

1. Seed samples
2. Forceps
3. Blade
4. Solution of tetrazolium chloride

Procedure

Soak the seeds in water. Give a cut to the seed wit sharp blade so as to cut the embryo. Put the cut seed in tetrazolium chloride solution. The embryo will turn pink if viable. Record your observations in following observation sheet:

OBSERVATION SHEET

Observations	Samples No.		
	1	2	3
No. of seeds taken for test			
No. of seeds showing pink colour of embryo			
No. of seeds not showing pink colour of embryo			

$$\text{Viability (per cent)} = \frac{\text{No. of seeds showing pink colour of embryo}}{\text{No. of seeds kept for test}} \times 100$$

Chapter 5
Nursery Raising Management

Vegetables are being grown by the vegetable growers to meet out the need of the family as well as the people of the country. Vegetable crops are raised either by seed or vegetative means. The crop propagated by seed are grown by direct seed sowing in the main field *i.e.* okra, peas, cowpea, French bean, dolichos bean, radish, turnip, carrot, bitter gourd, bottle gourd, ridge gourd, sponge gourd, ash gourd, watermelon, muskmelon, cucumber, pumpkin, round melon. Spinach beet, spinach, coriander, amaranth, lettuce, fenugreek etc. Seeds are sown in nursery beds to raise seedling and when seedling become ready for transplanting they are transplanted in seed bed (main filed) *i.e.*, tomato, brinjal, chillies, capsicum, cauliflower, cabbage, knoll-khol (kohl rabi), chinese cabbage, savoy cabbage, Brussels sprouts, Sprouting broccoli, endive, chicory (red and green), celery, fennel , sweet marjoram, kale, sorrel, mellow, salsify, parsley, lettuce etc. Some vegetables are propagated through vegetative means like colocaisia, yam, ginger, garlic, potato, sweet potato, drumstick, asparagus, water chestnut, nadroo, elephant foot, pointed gourds, momordica species and ivy gourd. But there are certain vegetables having very small seeds are first sown in the nursery for better care and to combat with the time for field preparation etc. and after about one month of seed sowing transplanted in the main field. There are many advantages of nursery raising.

It is very easy and convenient to look after the young tender seedlings growing in small areas against disease, insect, pests, weeds etc.

Favourable growing conditions can be provided easily for raising seedlings in nursery bed to protect them against bright sun, rains and low temperature (unfavourable whether conditions).

☆ By sowing the seeds in the nursery we can raise the early crop at least about 1.0 to 1.5 months and fetch higher price.

☆ By nursery sowing we can save land and labour so there is a great economy because seeds are sown in small area.

☆ More time is available for the preparation of main field because nursery is grown separately.

☆ Generally vegetable seeds are very expensive particular hybrids, so we can economies the seed by sowing them in the bursary.

Selection of Site

The following points should be considered while selecting nursery area:

☆ The area should be free from water logging

☆ It should be away from shade to get desired sun light

☆ The nursery area should be near the water supply

☆ The area should be fenced/protected from pet and wild animals

Soil and Soil Preparation

Soil for nursery raising should be loam to sandy loam, loose and friable, rich in organic matter and well drained. The soil pH should be close to neutral *i.e.* 7.0. Soil preparation needs a deep cultivation of the nursery land either by soil turning plough or by spade and subsequent 2-3 hoeing with cultivator. Remove all the clods, stones and weeds from the field and level the land. Mix 2 kg well rotten and fine farm yard manure/compost or leaf compost or 500 g vermicompost per square meter and mix in the soil. If the soil is heavy mix 2-3 kg sand per square meter so that the seed emergence may not be hampered.

Soil Treatment

For raising the healthy nursery soil must be treated with any means of the soil treatment to kill the harmful pathogens.

Treatment of Soil against Pathogens

Soil Solarization

During the high sunshine (from May-June), the temperature rises up to 45°C. During this time, transparent polythene of 200 gauge can be spread on the whole nursery area after wetting the soil for about 5-6 weeks. The margin of the polythene should be covered by wet soil (compressed mud) so that the air from in side could not come out from the covered area. After 5-6 weeks remove the polythene sheet and prepare the beds for seed sowing.

Formalin Treatment

The formalin solution should be done 15-20 days before seed sowing. First of all prepare formalin solution (1.5 to 2 per cent) in one container and drench the soil @ 4-5 litre of water per square meter soil surface so as to saturate it up to a depth of 15-20 cm. Cover the drench with polythene sheet of 200 gauge by putting the wet soil on the margin of the covered polythene sheet so as it does not allow the polythene film blown away by the wind and air from the covered area to outside. Removes the cover (polythene) after 15 days and prepare the beds for sowing. If there is insufficient time to wait for nursery sowing formalin dust can also be used. Prepare a mixture of formalin dust and well rotten FYM in the ratio of 15:85 mix well and spread the mixture on the beds @ 400-500 g per square meter. The spread material is mixed into the soil up to 5-6″ depth and then sow the seeds.

Application of Fungicides

There are certain fungicides like captan or thiram, which can also kill the soil borne pathogens. Use 5-6 g of any of the fungicides per square meter nursery area. For proper coverage, use dry soil (5 g fungicides per kg of soil). Spread the soil mixture on the nursery bed and mix well upto depth of 5-6 cm if at all there is some problem to broadcast captan or thiram, dissolve it in water @ 4-5 g/litre and drench the nursery bed @ one litre/sqm. When the nursery soil looks dry or friable, plough or dug the beds with spade and mix into the soil upto 5-6″ depth and prepare the beds for seed sowing.

Treatment of Soil against Insects

Application of Insecticides

There are certain insect pest and their eggs or secondary stage insects present in the soil and infest the seedlings in the later stage. To save the seedlings against them there are some insecticides also which are used as soil treatment to kill the unwanted pest present in the soil. The soil to be used for nursery raising should be treated with Furadan 3G@5 g/m^2 of the area. Furadan must be thoroughly mixed in soil up-to depth of 10-15 cm.

Steam Treatment

Steam can be used to treat the soil against harmful insect pest. First of all covered the required area with the help of polythene sheet and stop the air coming in and going out from the covered area. Supply the hot steam for at least 4-6 hours continuously. In this way all the harmful pathogen and insect pest may die. Plough the soil and prepare the beds for seed sowing.

Seed Treatment

The seeds are basic material to develop as plant. It also caries pathogen of certain disease and insect pest. To keep the seed free from these harmful diseases and pests seed treatment is very essential. Take the seeds to be treated in a wooden or glass jar having lid on it, apply captan or thiram @ 4g/kg seeds. Shake the jar well so that the seeds and chemical mix together well. Open the lid and use the seeds for sowing. Now a day most of the hybrid seeds and open pollinated varieties are sold in the market after the seed treatment. In such cases there is no need to treat them.

Nursery Bed Preparation

Nursery bed should be prepared according to the season and crop. In the rainy season, raised beds are prepared but in the winter and summer season either raised or flat beds should be prepared. For the uniform and high percentage of germination the soil must be fine and moist enough. If the seedlings are to be raised in boxes during un-favourable weather condition the flower pots, polythene bags, potting plugs, wooden trays, earthen pots etc may be used. Prepare soil mixture in seedling raising structure. Arrangement should be made to drain excess water from these structures by making a hole at the bottom.

Raised Nursery Beds

Depending on the level of the land, length of the bed may be kept 3 to 5 meter. However, width is restricted to 1 meter only which facilitates interculture operations. The beds are raised 15 to 20 cm high from the ground level. A space of 30-40 cm is left in between two beds. The beds are raised by utilizing the soil of this space and ultimately it converts into furrow.

The space between two rows helps in weeding, nursery care against diseases and insect pest and also for draining out the excess rain water from the nursery beds. The number of beds depends on the particular crop, season and growing area of the crop. The nursery bed should be smooth and slightly raised in centre compared to the margins so that water may not stagnate there and drained easily from the beds. The beds should be prepared in the east-west direction and line should be made north-south direction on the beds.

Flat Nursery Beds

Flat nursery method of nursery raising is suitable for spring and summer season. It is also suitable to raise the nursery in the places where the rainfall is very less soil is well drained, loose and friable. The nursery beds of the same size 3-5 m long and 1 meter wide should be prepared without raising the surface of the bed and in between the two beds 30-40 cm space should be left for weeding, application of plant protection measures, irrigation and other cultural works. In case of raising onion nursery a ridge between the two beds is prepared for light irrigation and other cultural works. The main drawback in this system of nursery raising is water stagnation during the rainy seasons.

Nursery Raising in Adverse Weather Condition

The vegetable seeds particularly the hybrids are very expensive therefore, it should be ensured that every seed produces a healthy seedlings and this is only possible if seedlings are raised under controlled conditions by maintaining 20-30°C temperature.

The seedlings can be raised in glass house, shade house or polyhouse in small structures like earthen pot, potting plugs, polyhtene bags and curd tray by using soil, sand and compost mixture peat/perlite.

During too low and too high temperature conditions the artificial temperature regimes can be created in the glass house or poly house for seed germination and raising the seedlings successfully to catch the early market by producing the early crop.

Requirement of Seeds and Seedling Area

The seeds and area for raising the seedlings may vary according to the soil crop, season and method of nursery raising. The details are given below in the Table 5.1.

Table 5.1: Requirement of Seeds and Seedling Area

Name of the Vegetables	Seed Rate (g)	Area Required (m²)
Tomato (Hybrid)	150-200	75-100
Tomato (OP)	250-300	100-125
Brinjal	300	150
Chillies	500-600	75-100
Capsicum	400-600	100-150
Early Cauliflower	700	150-200
Mid late Cauliflower	400-500	100-150
Cabbage	450-500	100-150
Knol-khol	700-750	200
Onion	8000-10000	500

Sowing of Seeds in the Nursery

After the seed bed preparation seeds are sown in the nursery bed either by broadcasting or in lines depending upon the nature and season of crops.

Broadcasting Method

In broadcasting method seeds are broadcasted on the well prepared nursery beds and later on the seeds are covered with mixture of well rotten fine sieved and treated FYM or compost, sand and soil in the ratio of 1:1. The major disadvantages of this method is uneven distribution of seeds in the nursery beds resulting some places vacant and some places have more number of seedling which result poor growth and development of seedlings. Sometimes seedlings become so dense and they look like as patches of grasses. In such cases there is more possibility of damping off disease occurrence.

Line Sowing

Line sowing is the best method of seed sowing in nursery. Lines are made 0.5 to 1.0 cm deep parallel to the width keeping 5 cm line to line distance. The seeds are sown or placed singly at a distance of about 1.0 cm apart. Cover the seeds with fine mixture of sand, soil and well rotten and sieved FYM or leaf compost etc. (1:1:1). After the seed covering a light irrigation must be given with the help of rose can.

Line sowing facilitates less incidence of damping off disease resulting to reduce seedling mortality. It has the following advantages also:

★ Each and every seedling will be healthy, bold and uniform.

★ Less seeds are required compared to the broadcasting method.

★ Every seedling will get uniform light and air.

★ Weed management of seed beds will be easy.

★ Seed bed covering mixture (soil: sand: FYM) will be required in small quantity as compared to the broadcasting method.

★ Easy to look after the seedlings if there is any disease or insect causing damage.

★ In hot weather condition thatched screen or shading nets are used to protect the nursery from the direct sunlight.

Seed Covering Material and its Treatment

After seed sowing either by broadcasting or line sowing method it is covered for better emergence. Therefore, a mixture of sand: soil: FYM in the ratio of 1:1:1 is prepared well mixed together and treated with any method of soil treatment as discussed above. It will be better to treat this mixture while treating the nursery soil. Apply 3-4 g thiram or captan per kg mixture. Care should be taken that every seed is well covered by seed covering material.

Use of Mulch

To maintain the soil moisture for seed germination cover the seed bed with a thin layer of mulch of paddy straw or sugarcane trash, or sarkanadas or any organic mulch during hot weather and by plastic mulch (plastic sheet) in cool weather. It has following advantages:

★ It maintains the soil moisture and temperature for before the seed germination.

★ It suppresses the weeds.

★ It gives protection from direct sunlight and raindrops.

★ Protects against bird damage.

Removal of Mulch

Due attention is given to remove the covered mulch from the seedbed. After three days observe the seed beds daily. As and when the white thread like structure is seen above the ground remove the mulch carefully to avoid any damage to emerging plumules. Always remove the mulch in evening hours to avoid harmful effect of bright sun on newly emerging seedlings. There are certain waiting durations in different vegetables are given in Table 5.2.

Name of the Vegetables	Seed Germination (days)
Tomato	5-7
Brinjal	5-6
Chillies	7-8
Cauliflower	3-4
Cabbage	3-4
Knol khol	3-4
Onion	7-10
Endive	3-4
Chicory	3-4
Savoy cabbage	4-5
Kale	3-4
Brussels sprouts	3-4
Celery	10-15
Chinese cabbage	3-4
Sorrel	8-10
Broccoli	3-5
Sweet Marjoram	10-12
Lettuce	4-5

Use of Shedding Net

After seed germination during the seedling growth, if there is very high temperature (>30°C) then beds should be covered by 50 per cent or 60 per cent shedding nets of green/green+ black coloured about 60-90 cm above the ground by the use of suitable support.

Watering

The nursery beds required light irrigation with the help of rose can till the seeds get germinated. Excess rainwater or irrigated water should be drained out from the field as and when it is required otherwise plants may die due to excess of water. Watering in the beds depends upon the weather conditions. If temperatures is high, open irrigation is applied. Need not to irrigate the beds during rainy days.

Thinning

It is an important operation to remove weak, unhealthy, diseased, insect pests damage and dense plants from the nursery beds. Keeping distance of about 1.0 cm from plant to plant. The thinning facilitates balance light and air to each and every plant. It also helps in watching the diseased and insect pest attacked plants while moving around the nursery.

Weed Control

Timely weeding in nursery is very important to get healthy seedling. If there are some weeds in the seed bed remove them manually either by hand or by hand hoe (thin forked khupri). Pre-emergence herbicides can also be sprayed soon after seed sowing to control the weeds. Stomp @ 3 ml/litre of water should be sprayed on the nursery beds after the seed sowing and seed covering with mixture of FYM soil and sand.

Nursery Raising of Direct Sown Crop

The direct sown crop can not be transplanted after uprooting because of the formation of an external layer of corky cell "Suberine" in the cell of roots. Since suberin is impervious to water resulting the death of the plants in the want of water. There are certain other combinations such as damage of growing tips and poor root regeneration which also hinder plant establishment after transplanting. But one can plant such vegetable crops with earth ball. Generally seedlings of direct sown crops are raised in the adverse situation when there is too low temperature for seed germination or the field preparation is late due to some standing crop in the field. Now there is practice in cucurbits to sow seeds in small poly bags (15 x10 cm size) filled with soil, FYM/compost and sand mixture (1:1:1). Sowing is done in the month of December in adverse weather when outside conditions are not favourable. Poly bags can kept in poly house where temperature is above 20°C germination. Watering is done regularly as per need. After seed germination growth takes place. As soon as outside conditions are favourable for planting, the seedlings are transplanted in well prepared and well fertilized pits either by removing the polythene or by cutting lower portion of the polythene so that roots may penetrate into the soil. Bottle gourd, bitter gourd, muskmelon, watermelon, cucumber etc may be grown in polyhouse and early crop can be raised. In this way we can advance our crop about 45 days than normal planting and also growing season of proceeding crops may be extended.

Nursery Raising of Vegetative Propagated Vegetables

There are certain vegetables which are neither propagated by seeds nor by seedlings in such case the stem and root cuttings are planted in nursery beds/polybags for mass multiplication.

In the vegetables like pointed gourd and ivy gourd, stem cuttings of one year old stems are used. The cuttings are planted in the polythene bags of 15 x10 cm size filled with mixture of sand, soil and well rotten FYM (1:1:1). The stem cuttings of pointed gourd 30 cm long turned in the shape of English '8' and buried in soil leaving 2-3 cm both ends exposed. While in the Ivy gourd 15 cm long and 2-3 cm thick stem cuttings are selected and planted in the polythene bags leaving 1.5-2.0 cm one side exposed. Irrigate the transplants with dose can regularly as and when required. In about one to two months root and shoot emerged from cutting. When the main field is ready the sprouted cutting are planted in the field either by removing polythene with earth ball or along with polythene by cutting bottom portion for penetration of roots in the soil. The best time for planting the cuttings in the bags is Oct. – Nov. in pointed gourd and Jan-Feb. in Ivy gourd. The cuttings become ready for transplanting in about 6 weeks after planting.

Hardening of the Plants in the Nursery

The term hardening includes any treatment that makes the tissues firm to endure better during unfavourable environment like low temperature, high temperature and hot dry wind etc. Hardening is physiological process whereby plants accumulate more carbohydrates reserves and produce additional quiticle on the leaves. In this process seedlings are given some artificial shocks atleast 7-10 days before uprooting and transplanting. Seedlings are exposed to the full sunlight all the shedding nets, polythene sheets should be removed and irrigation is stopped slowly and slowly.

Techniques of Hardening

The hardening is done by the following ways:

☆ By withholding the watering 4-5 days before transplanting.

☆ Lowering the temperature also retards the growth and adds to the hardening processes.

☆ By application of 4000 ppm NaCl with irrigation water or by spraying of 2000 ppm of cycocel.

Duration and Degree of Hardening

It is very necessary that plants should be hardened according to their kind. So that there is an assurance of high percentage of survival and slow growth under the condition to be expected at the time of transplanting. Hardening should be gradual to prevent or check the growth. Warm season crops like tomato, brinjal and chillies do not favour severe hardening. In Indian condition the hardening is done by allowing the soil become dry for 5-6 days.

Effect of Hardening

The following effect may be observed by the hardening:

☆ Hardening improves the quality and modifies the nature of colloids in the plant cell enabling them to resist the loss water.

☆ Hardening increase the presence of dry matter and retards the plants but decrease the percentage of freezable water and transpiration per unit area of leaf.

☆ Decreased the rate of growth in the plants.

☆ Hardened plants can withstand better against un-favourable wheather conditions like hot dry winds or low temperature.

☆ Hardening of the plants increase the waxy covering on the leaves of cabbage.

Plant Protection

Adaptation of plant protection measures in nursery against the incidence of insect pest and diseases is very important task to get the healthy seedlings. Among diseases damping off of seedlings, leaf curl, leaf blight, leaf miner, borer and leaf eating larvae infect the seedling in the nursery. Therefore timely care must be taken to control them.

Damping Off

This is very serious disease of nursery. Pre-emergence death of seeds or post emergence death of seedlings is observed. However, this disease mainly appears at 2-3 leaf stage. In first instance girdling takes place on the stem near base of the stem and seedlings bent down on the ground and die. This is a fungal disease mainly caused by *Rhizoctonia, Pythium, Phytophthora, Fusarium, Alternaria, Phomopsis* etc. Treat the nursery bed either by soil solarization, formalin solution or formalin dusts or fungicides like thiram or captan as discussed earlier. Treat the seeds as discussed in seed treatment. If the disease appear after the seed emergence, drench the nursery beds with 0.1 per cent solution of Brassicol or 0.7 per cent Captan or Thiram. It will be better to remove and buried the affected seedlings from the beds otherwise spread will be more.

The disease can be controlled to some extent by applying treated sand, soil and FYM mixture in the nursery bed up to the level from where the seedlings are falling.

Bio-agents for Control of Damping Off

Damping off disease in nursery can be effectively controlled by bio agents. Various species of *Trichoderma viz. Pseudomonas flouroscence* and *Aspergillus niger* can be used as soil treatment or seed inoculants. Bio agents can be used by two methods. It can be applied in well prepared and pulverized nursery bed @ 10-25 g/m^2 two days before sowing. Precaution should be taken to avoid sun or rain for 5-7 days after application. This can be also used for seed treatment. Bio agents containing *Trichoderma* should be well mixed with soil so that a thin layer of inoculation could be ensured over seeds. Bio-agents as seed inoculant are mixed @ 6.0 g/kg seed. After mixing seed should be dried in shade for one day. Approximately 10^8-10^{10} spores of solution is found effective for control of damping off. Following points should be considered before using bio-agents.

☆ Culture should be effective, viable and location specific.

☆ Nursery soil should be rich in organic matter.

☆ Soil should have sufficient moisture at the time of application.

☆ Nursery bed should be protected with sun and rain particularly in early stage.

Leaf Miner

It is very small insect enter in the leaves from margin side and move from one place to another by eating the chlorophyll. Initially the infected part of the leaves become brown and later on dry.

Control

☆ Spray 4 per cent neem seed kernel extract on the plants (crush 40 g of neem seed kernel, add some water and allow them for overnight. In the morning filter the extract and makeup the volume 1000 ml).

☆ Spray monocrotophos or metasystox 1.5 ml/ltr. water.

Raising of Virus Free Seedlings

Leaf curl is an white fly transmitted viral disease, infestation starts from seedling stage and continue till harvest of the crop. The disease is specially seen in the tomato and some time in chilli too and causes great loss of the crop. The leaves of affected plants show curling, mottling, rolling, puckering etc. It can be controlled by the following ways:

☆ Treat the soil of the nursery by Carbofuran @ $5g/m^2$.

☆ Seed treatment with Imidachloprid @ 2.5 g/kg seed.

☆ Cover the seed bed after sowing by Agronet making a tunnel like structure.

☆ Spray the nursery beds 15 days after seed germination at 7 days interval with Metasystox or Monocrotophos @ 1.5 ml/litre of water. Last spray is done 2 days before transplanting.

☆ Remove the infected plants if any in the field and buried in the soil or burn.

In this way the raised seedlings will be healthy and free from viral diseases.

Chapter 6
Nursery Raising Technology for Vegetable under Protected Conditions

It is said "A good seed sown in a good field results in a very good year". Production of healthy seedlings is the most important step of crop management to reap potential yield in vegetables. Healthy seedlings give the required quick start to crop establishment which results in optimum vegetative growth and help in realizing potential yield. In temperate regions vegetable seedling production is gradually changing from open field nurseries to protect raised bed or seedling tray production in some of the intensive vegetable growing areas. Increasing susceptibility of vegetables to various biotic and abiotic stresses and very high cost of hybrid seeds has warranted the attention of the vegetable growers to improve the nursery raising technology of vegetables. Now a days protected nursery raising of vegetables has become a full fledged industry in several developed countries like Israel, Japan, Spain, Netherlands, USA etc. In countries like Israel no vegetable grower is growing his own nursery but they are getting the required kind of seedlings from well established nurseries on pre-order basis system. Similar system of nursery raising is also prevalent in several other advance countries. In those situations where vegetables are being grown under protected conditions it becomes almost pre-requisite to raise the required vegetable seedlings only under protected environment to get virus free healthy seedlings.

Protective Structures for Seedling Production

Seedlings need care and nourishment and a protected enclosure is necessary to growth healthy and quality seedlings. Vegetable seedlings are being grown in low cost polyhouses, net houses, clotches/low tunnels, cold frames, Hot beds, lath houses etc. which provide control of growing conditions creating a micro environment congenial for propagation and cultivation of vegetable crops.

Hot Beds

The main objective of hot bed is to raise seedlings earlier and protect them from weather hazards. A hot bed is one where heat is generated by decomposition of fresh manure. The heat generated is utilized for seed germination which results in early nursery raising, early supply of vegetable produce in the market and more profits. First of all a trench 2 feet deep, 3 feet wide, 6 feet long is prepared. The frame generally made of wood and filled in such a way that from back side it extends up to 30-35 cm and from front side 20-25 cm above the ground. The sides of the frame are covered with paddy straw to prevent the loss of heat. The trench first filled with fresh manure upto 25-30 cm in two layers each separated with a layer of straw followed by 10-12 cm thick layer of straw followed by 10-12 cm thick layer of light soil. The top of the frame is filled with polythene lined lids used during night and rains.

Cloches/Low Tunnels

Cloches or low tunnels are also used for raising vegetable seedlings under unfavorable weather conditions. These cloches or tunnels are made curved and are covered with polythene. The end of theses cloches/tunnels can be closed with polythene sheets as per climatic requirement. Cloches prevent both hardening and frozening of land there by helps in sowing of seed earlier and when desired.

Thatches

Thatches are traditional structures used to protect the vegetable nurseries from unfavorable weather conditions both during winter and summer seasons. In winter thatches are erected in a slanting manner at 45° angle from ground level and are oriented in south west directions the slanting roof is covered with paddy straw or straw mats. The shade is removed when the seedlings have come up and have attained one centimeter height with two or four leaves.

Seed Panes/Boxes

Seed panes/boxes are used to raise delicate kind of seeds. Seed panes are shallow earthen pots about 4 inches high and 14 inches in diameter at the top with a single hole at the bottom. Seed boxes which are of wood 16 inches wide, 24 inches long and 3-4 inches deep with 6-8 holes drilled at bottom for effective drainage. Gravel stones or wood charcoal may also be put on the bottom of both panes and boxes to ensure proper and regular drainage. Then theses panes/boxes are filled with fine soil up to desired depth. Then the seeds are sown.

Polybag Nursery Raising

Nursery raising of cucurbits in polybags under protected structures is highly remunerative. Polybags of desired length (200 gauge, 20x10 cm size) taken and are perforated to ensure proper drainage. The polybag mixture consists of two parts garden soil, one part of sand+ one part FYM. The polybags are filled with this mixture upto desired height leaving some space empty. Then seeds are sown. Perforations are provided on all sides to ensure proper drainage and aeration.

Polyhouse/Net House

In a poly house the main frame can be of steel pipes and wooden poles of 6 to 8 feet high. It can also be erected with stone pillars to reduce the expenditure. Poly house are covered with 200 micron UV stabilized polyethylene film on the roof and the sides are covered with 40 mesh insect-proof nylon net. A refractable shade net is provided to bring down the temperature during summer days. In a net house stone pillars are erected as a main frame and the roof is generally covered with a shade net instead of polysheet. The sides are covered with the insect proof net. However, it is advisable to cover the roof also with the insect proof net above the retractable shade net to have better control over the entry of insect vectors like white flies. It is essential to harden the seedlings before transplanting. A retractable shade net will be use full to regulate the shade in the green house depending on the light levels. Plastic pipes of ¾ inch are bent in arch shape over the nursery beds and are covered with a plastic sheet to protect the seedlings from rain in a net house. The poly house and net house structure provides adequate light, shade and humidity. It protects the seedlings from thrips and white flies which spread viral diseases. Farmers can also grow vegetable seedlings in plastic trays on a small scale in their farms in a low cost net house measuring about 20 ft long, 10 ft wide and 8 ft high.

Based on the type of cladding material used in covering the installed structures. The poly house can be broadly divided into the following groups:

1. Fibre-glass polyhouse.
2. Single of double polyethylene film polyhouse
3. Ordinary glass house
4. Poly-carbonate house
5. Ultraviolet stable polyethylene film house

Seedling Trays

Seedling trays are also called as pro-trays (propagation tray) or flats, plug trays or jiffy trays. The most commonly used are 98 celled trays for Tomato, Capsicum, Cabbage, Cauliflower, Chilli. Brinjal and Bitter gourd. The dimension is 54 cm in length and 27 cm in breadth with a cavity depth of 4 cm. Trays are made of polypropylene and reusable. Life of the tray depends on the handling. Before using every time it is necessary that these trays are thoroughly washed and disinfected with a fungicide. The holes at the bottom of the cells control te moisture properly. Equal spaced cells facilitate equal growth of the seedlings. Seedling trays have been designed in such way that each seedling gets appropriate quantity of growing media and the right amount of moisture. Trays have pre pinched holes to each cavity for proper drainage of excess water and also have right spacing to facilitate equal growth of the seedlings.

Ingredients used as a Media for Growing Transplants

Well decomposed and sterilized medium is essential to grow disease free seedlings. The traditional potting mix of soil, manure, sand has been replaced over years by peat vermiculite, sand or Perlite mix. The most commonly used growing medium is coco peat and it retains optimum amount of moisture to support seed germination. Coco peat is a by product of coir industry and it has high water holding capacity. Neem cake (100 kg) and *Trichoderma* (1kg) are added per tone of the coco peat to prevent seedling diseases. Vermicompost is also used as a growing medium in place of coco peat. These ingredients are mixed in 3:1:1 ratio before filling the trays.

Organic Products

☆ Spagnum moss or peat moss containing 80-90 per cent organic matter 4-20 per cent of ash with a CEC of 60-120 meq/100g.

☆ Peat humus or brown peat containing 50 per cent organic matter 5.05 per cent ash with a CEC of 250-350 meq/100g.

☆ Organic wastes like saw dust, pine bark, pine chips, paddy husk and coir dust (coco peat).

Inorganic Products

Vermiculite

Holds and releases large quantities of water which rein force similar properties of peat when mixed with it. It is neutral in reaction and has relatively high CEC of 80 m eq/100g leaching of nutrients is reduced. The disadvantage is its high cost and early break down leading to compression of the substrates.

Perlite

Totally inert ahs low CEC or buffering capacity and low water content. It provides air space to the medium, neutral in pH, very light in weight and a good temperature stabilizer. Disadvantage 'Al' toxicity in some seedlings at low pH, limited capacity for water supply under conditions of high transpiration.

Sand

Porosity around 40 per cent of the bulk volume, particle size 0.5 to 2 mm in diameter contains no nutrient and has no buffering capacity. The CEC is 5.50 m eq/100g. It is used together with organic material.

Synthetic Products

Glass wool, polyurethane foam etc. can be used to grow seedlings.

The low pH of the substrate can be adjusted by adding lime (calcium carbonate) and dolomite (Ca -Mg) carbonate and with basic fertilizers like calcium nitrate, sodium or potassium nitrate. The high pH is adjusted by the adding of sulphur, gypsum, epsom salt and acidic fertilizers like urea, ammonium sulphate, ammonium nitrate, ammonium phosphate and acids like phosphoric acid and sulphuric acid.

Poor aeration of a highly decomposed black peat or of clay soil which on the other hand has an appreciable water retention capacity can be corrected by adding materials such as sand, polystyrene, perlite or expanded clay in which the common characteristics is to increase aeration. Sand and peat or coir dust mixture or peat and vermiculite or perlite supplemented with balanced fertilizers from the best medium to grow seedlings.

Disinfection of the Medium

Seedlings are very vulnerable to soil- born diseases and for seed and potting composts it is worth while sterilizing the soil before mixing the other ingredients of the compost. The soil of nursery beds (flat and raised beds) is disinfected by solarization or the beds and substrates of growing medium are sterilized with steam or formaldehyde to control soil/medium born diseases. Diseases such as Damping off can be arrested by chestnut compound. Pots and boxes should also be washed and dipped in a good disinfectant.

Methods of Seedling Raising

☆ Fill the seedling tray with the appropriate growing medium.

☆ Make a small depression (0.5 cm) with fingertip in the centre of the cell sowing. Alternatively depression can be created by stacking 10 trays one over other and pressing the trays together.

☆ Sow one seed per cell and cover with medium.

☆ No irrigation is required before and after sowing if coco peat having 300-400 percent moisture is used.

☆ Keep 10 trays one over other for 3-6 days depending on the crops. Cover the entire stack with polyethylene sheet. This ensures conservation of moisture until germination. No irrigation is required till seedling emergence. Care must be taken for spreading the trays when the seedlings is just emerging otherwise seedlings will get etiolated.

☆ Seeds start emerging after about 3-6 days of sowing depending upon the crops. Shift the trays to poly or net house and spread over a bed covered with polyethylene sheet.

☆ The trays should be irrigated lightly every day depending upon the prevailing weather conditions by using a fine sprinkling rose can or with hose pipe fitted with rose. Never over irrigate trays as it results in leaching of nutrients and building up of diseases.

☆ Drench the trays with fungicide as a precautionary measure against seedling mortality.

☆ The media may need supplementation of nutrients if the seedlings show deficiency symptoms. Spray 0.3 percent (3g/litre) of 100 percent water soluble fertilizer (19 all with trace elements) twice (12 and 20 days after sowing).

☆ Protect the trays from rain by covering with polyethylene sheets in the form of low tunnel.

Polyhouse

Tunnels or Row Covers

Thatches

Plastic Trays

Polybags

Hot Bed

Raised Beds

Cloches

Figure 6.1: Different Protected Structures Used for Early Nursery Raising.

Harden the seedlings by without holding irrigation and reducing the shade for a week before transplanting.

The seedlings will be ready in about 21–42 days for transplanting to the main field depending upon the crops.

Advantages of Seedlings Production in Tray

☆ Seeds germinate properly and mortality of seedlings is negligible.

☆ No loss of expensive seeds of hybrids.

☆ Adequate space for each seedling to grow properly.

☆ Damage due to pests and diseases is very rare.

☆ Promotes better root growth.

☆ Transplanting shock is negligible.

☆ Easy to handle and transport

☆ Seedlings do not wither during transport.

☆ Uniform growth in nursery ensures better establishment and growth of plants in the main field.

Chapter 7
Water Management

Irrigation

Irrigation is the artificial application of water to soil for the purpose of vegetable production. Irrigation water is supplied to supplement the water available from rainfall and the contribution from ground water. Generally the timings of rainfall are not adequate to meet vegetable crop water requirement and irrigation is therefore, essential to raise successful crops.

Need of Irrigation

Irrigation is needed for one, more or all of the following:

☆ To add water to soil to supply the moisture essential for plant growth.

☆ To provide crop insurance against short duration drought.

☆ To cool the soil and atmosphere, thereby making more favourable environment for growth.

☆ To wash out or dilute salts in the soil.

☆ To reduce the hazard of soil cracks

☆ To soften tillage pans.

Water Requirement and Irrigation Efficiency

The estimation of the water requirement (WR) of crops is one of the basic needs for vegetable crop planning on the farm and for the planting of irrigation projects. Water requirement may be defined as the quantity of water, regardless of source, required by a crop or diversified pattern of crops in a given period for its normal growth under field conditions at a place. Water requirement includes the losses due to evapo-transpiration (ET) or consumptive use (Cu) plus the losses during the application of irrigation water (unavoidable losses) and the quantity of water required for special purpose such as land preparation, transplanting, leaching etc.

It may thus be formulated as:

WR = E T or Cu + application losses + special needs.

Water requirement is, therefore, a 'demand' and the 'supply' would consist of contribution from any of the sources of water, the major source being the irrigation water (IR), effective rainfall (ER) and soil profile contributions (S). Numerically, therefore, water requirement is given as:

WR = IR + ER + S

The field irrigation requirement of a crop, therefore, refers to the water requirement of vegetables, exclusive of effective rainfall and contribution from soil profile and it may be given as:

IR = WR – (ER + S)

The farm irrigation requirement depends on the irrigation needs of individual vegetable crops, their area and the losses in the farm water distribution system mainly by way of seepage.

Water Application Methods

Irrigation may be applied by flooding on the field surface, by applying it beneath the soil surface, by spraying it under pressure or by applying it in drops. The water supply, the type of soil, the topography of the land and the crop to be irrigated determine the method of irrigation to be used. Commonly, following methods are followed for irrigating different vegetables.

Border Irrigation

The border method of irrigation makes use of parallel ridges to guide a sheet of flowing water as it moves from slope. The land is divided into long, narrow and parallel strips called the borders that are separated by shallow ridges. The border strip has a uniform gentle slope in the direction of flow. The essential feature of border is to provide an even surface over which the water can flow from the slope with a uniform depth. Each strip is irrigated uniformly and independently by turning in a stream of water at the upper end. The border method is suitable to irrigate all close growing crops like okra, peas, beans etc. This method has number of merits:

☆ Border ridge can be constructed economically with simple farm implements.

☆ Labour requirement in irrigation is greatly reduced as compared to other methods.

☆ Uniform distribution of high water efficiency is achieved if the system is properly designed.

☆ Large irrigation streams can be effectively and easy.

☆ Adequate surface drainage is provided. If outlets are available.

Limitations

☆ It requires relatively high flow of water (Larger streams).

☆ It needs extensive grading of land.

Check Basin Irrigation

Check basin irrigation is the most common method of irrigation in India. It consists of dividing the field into smaller unit areas so that each has a nearly level surface. Bunds or ridges are constructed around the area forming basins within which the irrigation water can be controlled. The basins are filled to desired depth and the water is retained until it infiltrates into the soil.

Check basin methods is most suited to smooth gentle and uniform land slopes and for soils having moderate to slow infiltration rates.

Advantages

☆ Weeds are killed due to water standing for some time.

☆ It is also suitable for lands with gentle slope.

☆ Large layers of fertile silt are deposited in the basins.

Limitations

☆ Ridges of basin interfere with the movement of animal or tractor drawn implements.

☆ Considerable land is wasted under the ridges and bunds.

☆ Precise land grading and leveling is required.

☆ Labour requirement is fairly high.

☆ Method is not suitable to those crops, which are sensitive to wet soil conditions.

Furrow Irrigation

Furrow method of irrigation is used in the irrigation of row crops with furrow developed between the crop row in the planting and cultivating processes. The size and shape of the furrows depend on the crops grown, equipment used and spacing between crop rows. Water infiltrates into the soil and spreads laterally to irrigate the areas between the furrows.

Merits

☆ This method is used to irrigate all cultivated crops planted in rows, including orchards and vegetable crops.

☆ It is suited to most soils except sandy soils.

The fear of soil erosion is reduced with this method on wide range of slopy soils by carrying a furrow across the slope.

Ring or Basin Irrigation

The ring method of irrigation is essentially the check basin method especially adapted to irrigate the Cucurbits and tuber crops. The shape of basin may be circular, rectangular or square. Generally a ring is made for each plant, but under certain conditions of soil and surface slopes (2.5 per cent or more). The irrigation efficiency is about 50-60 percent.

Advantages

- ☆ Large streams of water can be utilized.
- ☆ Irrigation water can be controlled nicely.
- ☆ Water can be distributed uniformally.
- ☆ High water application efficiency can be achieved.
- ☆ Maintenance cost is low.
- ☆ Soil erosion is also controlled.

Limitations

- ☆ Involvement of labour is higher.
- ☆ Higher flow of water is needed.
- ☆ More wastage of land under bunds.

Sprinkler Irrigation

It consists of supplying water through nozzles or perforated pipes in the form of spray. The perforated pipes may be fixed on the side of a drum mounted on a truck or may be stationary. The stationer system may be permanent or temporary. The perforated pipe may be revolving on vertical stands and either fed through hoses or be permanently connected to pipes buried in the ground. Such revolving sprinklers serve comparatively larger areas. The whole system can be promptly and effectively control the water supply from the source. The sprinkler irrigation offers a better and more efficient control of water than the traditional surface methods.

Recognition of importance of sprinkler irrigation is an integral practice of modern agriculture. With proper fertilizers, insect and plant disease control, improved seed strains and mechanical farming operations, the sprinkler irrigation will provide the essential moisture control needed for optimum vegetable crops yield.

Advantages of Sprinkler

- ☆ Increased efficiency of water application leading to reduced gross water requirement.
- ☆ Improved soil moisture control for certain vegetables.
- ☆ Elimination or reduction of water logging and saving in cost of subsequent drainage.
- ☆ Elimination or erosion by irrigation water.
- ☆ Better weed control.

☆ Better distribution of water over the entire field.

☆ Less labour is required for irrigation

☆ Land leveling is not so much necessary.

☆ There is no crust formation on the soil.

☆ Reduction in losses and saving water.

Drip Irrigation

Drip or trickle irrigation is one of the latest methods of irrigation which is becoming increasingly popular in areas with water scarcity and salt problems. It is a method of watering plants frequently and with a volume of water approaching and consumptive use of the plants, thereby minimzing conventional losses as deep percolation, runoff and evaporation. In this method, irrigation is accomplished by using small diameter plastic, lateral lines (pipes) with devices called "emitters" or "drippers" at selected spacing near the base of the plants. The system applies water slowly to keep the soil moisture within the decided range for plant growth.

Advantages

☆ Crops like tomato, capsicum, chilli, brinjal etc. respond well to drip irrigation.

☆ Considerable saving in water by adopting this method is achieved, since the water is applied more precisely in the root zone only.

☆ Substantial increase in yields of vegetable crops has been observed by adopting this method.

☆ Reduces the concentration of salts in the root zone when irrigated with poor quality water.

☆ Like sprinkler method, drip irrigation permits the application of fertilizers through the system.

☆ It achieves 90 percent water application efficiency.

Limitations

The initial cost of drip irrigation equipments is considerably high for large scale adoption *i.e.,* economic considerations limit its use only in the vegetable crops in water scarcity areas.

Sub-surface Irrigation

This consists of supplying water to the soil by means of a network of concrete, wooden or stoneware pipes with open joints or special orifices and laid 0.4 to 1 m below the surface at a distance of 1 to 3 m under the entire area to be cultivated with the objective of economizing the quantity of water.

Advantages

☆ There is no loss of water by surface evaporation and a considerable economy in total consumption of water is affected. Usually $1/4^{th}$ to $1/8^{th}$ of the water required for surface flow irrigation alone would be necessary.

☆ It does not create crust formation on the surface and hence saves the labour of frequent cultivation for preserving tilth.

Limitations

☆ It involves a huge initial expenditure and therefore, can be used only for high value vegetables.

☆ The fibrous root system of plants may block the pores.

☆ It requires a considerable care in laying the pipelines which should be level for an even distribution of water.

Scheduling of Irrigation

From practical point of view, the major aspects of irrigation management are when and how much to irrigate the crops? During the recent years, several efforts have been made by researchers to develop suitable criteria for scheduling irrigation to crops based on crop, soil, atmosphere and plant water relations. The most common approaches are:

(1) Physiological Stages Approach

Some physiological stages of crop growth are found to be more critical in their demand for water than other stages when water supply is limited. It is necessary to take into consideration the critical stages of crop growth for scheduling irrigation.

(2) Soil Water Regime Approach

The criteria of scheduling irrigation on the basis of optimum fixed time interval between successive irrigations was dropped because both the threshold soil-water content and the fixed interval for the same crop varied with soil texture. Under this concept the water content at field capacity is taken as 100 percent available for crop plant and that at permanent wilting point as zero percent available. The safe limits of allowable soil water depletion for different crops are determined by field experimentation and taken as a criterion for scheduling of irrigation to crops.

(3) Transpiration Ratio Approach

Transpiration ratio is the amount of water transpired by a crop in its growth period to produce a unit weight of dry matter. In this method, plants are grown in pots and the evaporation is minimized by sealing the soil surface. The transpiration of the potted plants is considered as their water requirement. Now it proved that the transpiration and dry matter production by a plant are not related in a simple manner as in the transpiration ratio. The growth of various plants as influenced by several factors like soil, plants and atmosphere, therefore, transpiration ratio is of the little value in irrigation management under field conditions. Thus this approach is no more in use.

(4) Irrigation Depth and Interval of Irrigation Approach

In this approach arbitrary depth of irrigation and interval of irrigation are used for scheduling irrigation. The soil status at the time of irrigation and the rainfall received during the irrigation period is not taken into consideration. Therefore, this approach is improved as unscientific for scheduling irrigation to crops.

(5) Evaporation Approach (IW/CPE)

The climatic factors play an important role in deciding the water needs of crops and the criteria of soil water regime for scheduling of irrigation cannot be considered without taking into account the climatic factors. This leads to the concept of evaporation which is now used as criteria for scheduling irrigation.

Pan evaporation is used to estimate the amount of water given by the ratio of irrigation water (IW) to cumulative pan evaporation (CPE). For example, if irrigation is scheduled at 0.8 IW/CPE ratio and the depth of irrigation is 6.5 cm then irrigation would be given when the CPE reaches 8.12 cm. More practical criteria for scheduling irrigation *i.e.* IW/Epan. The ratio between a known depth of irrigation minus rains since previous irrigation (Epan).

Machines Used in Irrigation

Human and bullock power lifts are suitable for irrigation small areas when the supply of water in wells is limited. When the irrigated area exceeds 8 acres and water is at a great depth in the well, pumps operated by oil engine or electricity are more convenient and economical.

Oil Engine and Pumps

Oil engine with 3 to 10 HP are ordinarily used for irrigation with 2 to 4 inch pumps and ½ to 1 acres are irrigated every day. The equipment is, however, costly.

Electric Motors and Pumps

Electric motors with 3 to 10 HP are commonly used for irrigation with 2-4 inch pumps. The installation is not as costly as the oil engine pumping units.

Centrifugal Pumps

These are commonly used with oil engines and electric motors. The pump consists of an airtight casing in which is enclosed a flat impeller with rotating vanes. The impeller rotates at 1200 to 1500 revolutions per minute and sucks water when connected to a suction hose and discharge it through the delivery end. The pumps work efficiently up to a depth of about 22 feet and efficiency is reduced when the depth exceeds 25 feet. The power required for operation depends upon the total height over which water is lifted and the quantity so raised. Ordinarily 10 HP motors are used for lifting water over 30 to 50 feet with pumps having 4-inch sucker.

Measurement and Calculation of Irrigation

The irrigation water is measured under two conditions *viz.*, at rest and in motion. The units commonly used for expressing the volume of water at rest are litres, cubic meters, hectare meters, etc. The rate of flow is expressed in terms of volume per unit time, *e.g.* litres per second, cubic meters per minute, hectare meter per hour.

Chapter 8
Nutrient Management

Introduction

Nutrition to vegetable crops can be provided by application of organic matter and by chemical fertilizers. Organic matter not only forms a very important source of plant nutrients but it also affects certain properties of the soil. It is clear that many improvements are possible in fertilizer efficiency. These include improved water management and it increases the water holding capacity of the soil and improves its structure, efficient crop combinations, and improved formulations such as slow release fertilizers to synchronize nutrient availability with crop needs. The timing of fertilizer application is also important, not only to minimize leaching losses, but also to maximize yield. Vegetables are very sensitive to nutrient deficiencies during the early growth period. The addition of organic matter through the debries of the vegetable crops is very little and quite insufficient to cause any beneficial effect. For vegetable production both organic and inorganic manure are in use but inorganic *i.e.* chemical fertilizers provide nutrients very quickly.

Plants require 16 elements for normal growth and reproduction. It is generally agreed today that an ideal agricultural system must be not only productive but also sustainable. Some types of vegetable production are presumably more sustainable than others, but to define which these are we need indicators of sustainability. It is still difficult to define these indicators especially in quantitative terms. General indicators of sustainability have been suggested, including productivity (yield), stability (yield over time), protection of the resource base, viability (whether a system is economically profitable) and social acceptability (to both farmers and the general public). The problem is to apply such criteria to give quantified results.

Because of the high levels of nutrients removed with the crop, and because vegetables are high value crop, vegetable often receive very heavy fertilizer applications. Unbalanced fertilizer use and micronutrients deficiencies are common problems. There is a wide spread need among the regions vegetable farms for more soil testing, including perhaps a mobile service specially adapted to the needs of small-scale farmers. The crop nutrient requirement for a particular element is defined as the total amount in kg/ha of that element needed by crop to produce economic optimum yield. This concept of economic optimum yields is important particularly for vegetables.

Fertilizer

"Any natural or manufactured material, dry or liquid added to soil in order to supply one or more plant nutrients other than lime or gypsum is known as fertilizer."

Fertilizers and manures are used to supplement the nutrients required by the plants from soil to increase the crop yield *vis-à-vis* to maintain/improve the soil fertility. Continuous cropping and several other facts that necessitate the use of manures and fertilizers. The dose method and time of application depend upon crop, soil, fertilizer/manure and climatic factors of the region.

Principles of Fertilizer Application

The basic principle of fertilizer application is to make the nutrients readily available to the plants as their requirement without much wastage and harmful effects on soil. Usually larger quantities of fertilizers are added to clayey soils at longer intervals than to sandy soils because clayey soils are riches in humus than sandy soils and both clay and humus have a high capacity to retain nutrient ions this phenomenon called base exchange. These adsorbed nutrient ions are not lost by leaching cannot gradually taken up by the plant roots. If a heavy dose of water-soluble fertilizer is applied to sandy soil most of it will be leached down by high rainfall in the humid regions.

Quantities of Fertilizers to be Applied

Cultivated soils, where vegetable crops are raised have been placed in low or medium or high category according to their content of available N, or P or K. If the soils are low in one or more of the concerned nutrients, application of the nutrients in a full dose to the soils will increase the yield. If the soils are in medium category, only half the dose of that nutrient is applied to the soils. Different crops required different quantities of nutrients *e.g.* dwarf variety of tomato okra etc require high dose of nutrients then tall varieties, because they will lodge if supplied with high dose of nutrients.

Time of Application of Fertilizers

The rate of assimilation of N by vegetable crop is equal to the rate of growth. Vegetables require less N immediately after germination because they grow less at that time. The vegetable crops demand for N increase from the early growing stage to the flowering stage, when their growth rate increases. Therefore, nitrogen fertilizers should be applied in split doses. The vegetable crops utilize about two thirds of their phosphatic and potash requirement during their early growth period. So the entire quantity of P and K should be applied as basal dose.

Kind of Fertilizer to be Applied

Nitrate nitrogen should not be applied to sandy soils, especially under conditions of high rainfall because they are readily leached. Therefore, nitrate fertilizers are usually not applied to leguminous vegetables, where ammonical amide fertilizers (urea) should be applied.

Methods of Application

Broadcast, localized placement and spraying of fertilizer solution in vegetables are the three main methods of fertilizer application.

Broadcast

Broadcast of fertilizers means the uniform spreading of fertilizer over the entire field. It proves effective when heavy doses of nitrogenus and potassic fertilizers are to be applied to the acidic soils. In India, cultivators broadcast the fertilizers on the field and incorporate them into the soil by ploughing. Therefore, the field is thoroughly prepared and seeds of the vegetables are sown. Broadcast of fertilizers (usually nitrogenous) over the standing vegetable crops is known as top dressing.

The only advantage of broadcast is that fertilizer may be more conveniently applied in the hilly regions by this method. The broadcast of fertilizers should be discouraged due to following disadvantages:

☆ Most of the plant nutrients are assimilated by weeds.

☆ Fertilizers come in contact with large volume of soil, so most of the nutrients, especially phosphates, are fixed up.

☆ Nutrients cannot be fully assimilated by the roots of widely spaced crops.

Localized Placement

☆ The localized placement of fertilizers means their application very near the seed or the plant.

Drill Placement

Drill placement of fertilizers means the placement of the fertilizer in the soil with the help of drills, which usually have separate seed box and fertilizer hopper with drills attached and that seed and fertilizer are put in same furrow. The fertilizers if placed a little below the seed, will not harm the seed and the tender roots of the young seedling. Drill application saves cost of the labour.

Plough Sole Placement

Plough sole placement means the application of fertilizer in a continuous band at the bottom of the furrow when the field is being ploughed. This method encourages the development of a deeper root system, because fertilizers have been placed in the moist sub-soil, where roots develop.

Band Placement

The fertilizer is placed by the side of the crop, either along the row of crop or in the hill near each plant. This is called band placement, where the fertilizers are placed to one side or both sides of row of the vegetable plants at a distance of 2.5 to 7.5 cm from it. As such, band placement is of two types: Hill placement and row placement.

(1) Hill Placement

When the plants are spaced 90 cm or more, fertilizers are placed close to the plant in bands on one or both sides of the plants. The length and the depth of the bands and their distance from the plants vary with the crop and the amount of fertilizers. Hill placeent is practiced for N and P fertilizers to vegetables like pointed gourd, other cucurbits and tuber vegetables.

(2) Row Placement

When the seed or plants are sown close together in a row, the fertilizer is put in continuous bands on one or both sides of the row by hand or seed drill. This method of application is known as row placement. This method is practiced in potato, tuber crops etc. Higher rates of application are possible with row placement than with hill placement. However, for applying small amounts of fertilizers, hill placement is usually effective.

Pellet Application

In this localized placement method, the nitrogenous fertilizers are applied in the form of pellets 3 to 6 cm deep between the rows of the vegetable crop and it is known as pellet application.

Side Dressing

In this method, fertilizers are spread in between the rows or around the plants. As such side dressing is the broad term covering various practices of applying fertilizer. The common methods of side dressing are as follows:

   ☆ Application of nitrogenous fertilizers in between the rows by hand to broad row crops like potato, sweet corn, tuber crops etc.

   ☆ Application of mixed or straight fertilizer around the base of the vegetable crops like pointed gourd, ivy gourd, elephant foot, yam etc.

Type of Fertilizers

In India, fertilizers consumed are of following types: Nitrogenous fertilizers, phosphatic fertilizers, potassic straight fertilizers, complex fertilizers and fertilizer mixtures.

Nitrogenous Fertilizers

These fertilizers supply nitrogen. The common nitrogenous fertilizers are ammonium sulphate calcium ammonium nitrate and urea etc. Ammonium sulphate and urea are by far the most important nitrogenous fertilizers used by Indian farmers. The nitrogen content of some nitrogenous fertilizers is given below:

Name of the Fertilizer	N Content (Per cent)
Ammonium sulphate	20.6
Urea	46.0
Calcium ammonium nitrate	25.0
Ammonium sulphate nitrate	20.6
Ammonium chloride	25.0
Anhydrous ammonia	82.0
Ammonium nitrate	32.5

Urea

Urea is the most concentrated form of the solid nitrogenous fertilizers containing 46 percent nitrogen.

☆ It is white crystalline, organic chemical

☆ It is fairly hygroscopic in nature

☆ It is soluble in water

☆ It has less tendencies to stick and cake than ammonium nitrate

☆ It lacks sensitive to fire and explosion

☆ It has less corrosiveness to handling and application equipments

Ammonium Sulphate

This is one of the oldest sources of ammonium nitrogen.

☆ It has low hygroscopicity

☆ It has greater chemical stability

☆ It has better agronomic suitability

☆ It is white to yellowish grey in colour

☆ It contains about 20.5 per cent nitrogen and 23.5 per cent sulpur

Ammonium Nitrate

It is quite hygroscopic in nature and care must be taken in handling and storage for reducing/preventing caking.

☆ It has risk of explosiveness

☆ It is more prone to leaching and denitrification than ammonical products

☆ It contains 35 per cent N both in ammonical and nitrate forms

☆ It is dangerous in pure form because of explosion hazards

Ammonium Chloride

This fertilizer usually contains 25 per cent nitrogen. About two-third of the world capacity for manufacturing of this material is located in Japan with the remaining one-third situated in India. It has higher N content than ammonium sulphate.

Calcium Ammonium Nitrate

Calcium Ammonium Nitrate is brown or light grey in colour. It contains 25 per cent N, one fourth of which is ammonical and three-fourth is in nitrate form. It is available in granular form.

Calcium Nitrate

It contains approximately 15 per cent N, and is actually a complex double salt of calcium and ammonium nitrates. It is sold in prilled form. It is readily available to plants as nitrogen is in nitrate form.

Phosphatic Fertilizers

Phosphatic fertilizers are chemical substances that contain nutrient phosphorous in absorbable form. The primary material of phosphatic fertilizers is rock phosphate. The commonly used phosphatic fertilizers and their characteristics are given below:

Single Super Phosphate

It is ash coloured powder like material containing about 16 per cent P_2O_5, 20 per cent calcium and 12 per cent sulphur.

Di-calcium Phosphate

It contains 32-36 per cent P_2O_5 and is chemically intermediate between the water-soluble mono calcium phosphate and the insoluble tricalcium phosphate, which occurs in rock phosphate. It is slightly soluble in water.

Tri-calcium Phosphate

It contains 46-48 per cent P_2O_5. It is also known as triple super phosphate or concentrated super phosphate.

Bone Meal

Bone meal have been used as manures for time immemorial. Bone meals are of two types: (1) Raw bone meal (2)Steamed bone meal. Raw bone meal contains about 25 per cent P_2O_5 and 4 per cent N which is in the slow acting organic form. Steam bone meal contains 25-30 per cent total phosphorus (P_2O_5) and about 1-2 per cent N. It contains about 25 per cent citrate soluble phosphorus (P_2O_5); Steam bone meal is applied to soil few days before sowing of the crops.

Basic Slag

Basic slag is a by product of the steel industry, where the original iron ores contain appreciable amount of phosphorus. It is grayish black powder with a very high specific gravity. It contains 8-12 per cent P_2O_5.

Potassic Fertilizers

Potassic fertilizers are applied to soil supply the plant with potassium (K), one of the essential elements for plant growth. Main potassic fertilizers which are used today are:

Murate of Potash

It is a coarse or a fine salt resembling ordinary salt and having no bitter taste. It contains 60 per cent K_2O and it is reddish or dirty white crystalline material.

Sulphate of Potash

Sulphate of potash is a main non-chloride form of potash fertilizer. It is dirty white powdery material, which contains 48 to 52 per cent K_2O.

Complex Fertilizers

Straight Fertilizers vs Complex Fertilizers

Straight fertilizers supply only one of the primary fertilizer elements, either N or P or K for plant growth *e.g.* urea. Complex fertilizers supply more than one fertilizer elements needed for vegetable growth. When they supply any of the two of the fertilizer elements needed for plant growth, they are called incomplete fertilizer *e.g.*, Diammonium phosphate. When they supply all the three fertilizer elements for vegetable crops growth, they are called complete complex fertilizer, *e.g.* Nitro phosphate.

Mono and Diammonium Phosphate

Mono ammonium phosphate as well as diammonium phosphate is soluble in water. Mono ammonium phosphate is a light grey material, whereas diammonium phosphate is a dark brown material. Then absorb a little moisture from the atmosphere and are acidic in nature. Mono ammonium phosphate contains 11.0 per cent N and 48.0 per cent P_2O_5, whereas diammonium phosphate contains 18 per cent N and 46 per cent P_2O_5.

Nitrophosphate

Nitrophosphate is a granulated fertilizer-containing stabilizer, which prevents reversion of citrate soluble phosphate to insoluble phosphate. Being granulated, it maintains excellent physical condition during storage and handling. Nitrophosphate contains nitrogen and phosphorus, the major plant nutrients of various proportions depending upon the type of process of manufacture. If it is necessary, potassium salt are also added to it. Three types of Nitrophospates- 20:20:2, 18:18:9 and 15:15:15 are manufactured.

First type contains 20 per cent N, 20 per cent P_2O_5 and 2 per cent K_2O.

Second type contains 18 per cent N, 18 per cent P_2O_5 and 9 per cent K_2O

Third type contains 15 per cent N, 15 per cent P_2O_5 and 15 per cent K_2O

Usually about one third of the phosphate contained in these is water soluble and remaining two third is citrate soluble. The water soluble phosphate is readily available and hence useful to the early stage of vegetable crops growth, while the citrate soluble phosphate helps the crop to grow and attain maturity.

Mixed Fertilizers

A mixed fertilizer is a mixture of more than two straight fertilizers, *e.g.* ammonium sulphate and single super phosphate may be thoroughly mixed to get a mixed fertilizer.

Advantage of Mixed Fertilizer

☆ Two or more fertilizer elements are added together to make a mixed fertilizer to be applied in the field. Less labour is, therefore required for application of a mixed fertilizer.

☆ Fertilizer elements can be more uniformly applied to the field especially when they are required in small quantities.

☆ Mixed fertilizer can easily be drilled because of good physical condition.

Disadvantages of Mixed Fertilizer

☆ The use of mixed fertilizer does not permit the use of single nutrient which may be required by the crop at a certain stage.

☆ The illiterate farmers cannot effectively control the quantity of plant food nutrients present in the mixture.

Types of Mixed Fertilizers

Mixed fertilizers are of two types:

Open Formula Mixture and Close Formula Mixture

In open formula mixed fertilizer, manufacturer discloses the name and quantities of the straight fertilizers that are constituents of the mixed fertilizer, whereas in close formula mixed fertilizer, firm does not disclose the constituents of the fertilizer.

Sulphur Containing Fertilizers

These are the chemical substances containing the nutrient 'S' in the nutrient form of absorbable sulphate anions (SO_4^{-2}). The important 'S' containing water-soluble fertilizers are:

Sl.No.	Name of the Fertilizer	S content (per cent)
1.	Ammonium sulphate	24
2.	Potassium sulphate	18
3.	Gypsum	18
4.	Ammonium sulphate nitrate	15
5.	Single super phosphate	12

Micronutrient Fertilizers

Iron Fertilizers

These are generally water-soluble substances predominantly sprayed as foliar nutrients on the vegetable crops. Plants absorb iron in the form of Fe^{-2}. Commonly used iron fertilizer is Ferrous sulphate ($FeSO_4.7H_2O$) which contains 20 per cent Fe (water soluble).

Manganese Fertilizers

Manganese sulphate ($MnSO_4.4H_2O$) is pink salt containing 24 per cent Mn. Suitable for foliar application.

Zinc Fertilizers

Zinc sulphate ($ZnSO_4.7H_2O$) is water soluble salt, whitish in colour containing 23 per cent Zn. Can be applied as foliar or in soil.

Boron Fertilizers

Borax ($Na_2BO_4.10H_2O$) is water soluble white salt, which can be applied as soil dressing/foliar spray. It contains 11 per cent boron.

Other micronutrients like Cu and Mo are supplied through copper sulphate and sodium molybdate, respectively.

Practice to Learn

Identify the given samples of fertilizers and record the observations in the observation sheet.

OBSERVATION SHEET

Sample No.	Name of the Fertilizer	Type of Fertilizer	Nutrient Content	Specific Identifying Characters like Colour, Texture etc.
1.				
2.				
3.				
4.				
5.				
6.				
7.				
8.				

Practice for Learning

Acquaintance with methods of fertilizer application.

A. Fertilizer Application Practice at/before Vegetable Crop Sowing

Procedure

1. Calculate the amount of different fertilizers for given area on the basis or recommendations for the given vegetable crops.

2. Apply the fertilizers using different methods.

3. Record details of your observations/experience in the following observation sheet.

OBSERVATION SHEET

Item	Observations
Vegetable crops	
Recommended doses of fertilizers	
Schedule of application	
Area allotted for practice	
Sources of fertilizer available	
Quantity of fertilizers for the allotted area	
Methods followed (a) (b) (c) (d)	
Advantages of each method (a) (b) (c) (d)	
Problems in each method (a) (b) (c) (d)	

(b) Fertilizer Application Practices in Standing Vegetable Crops

Procedure

1. Calculate the amount of fertilizer for the given area on the basis of recommendations.

2. Apply the fertilizer using different methods applicable to the given vegetable crops.

3. Record details of your observations/experience in the following observation sheet.

OBSERVATION SHEET

Item	Observations
Crop	
Recommended dose of fertilizers	
Schedule of application	
Area allotted for practices	

Item	Observations
Source of fertilizer to be applied	
Quantity for the allotted area	
Methods followed	
Moisture condition of soil	
Stage of vegetable crops growth	
Precautions required	
Any other observations	

For proper growth and development of vegetable crop plants, all the essential nutrient elements are not required in optimum quantities. Some are required in relatively larger quantity while others in small quantity. These requirements depend on a number of factors *viz.*, vegetable variety, yield level, soil characteristics and environmental factors. Fertilizers and manures are used to supplement the nutrients required by the vegetables. Based on the series of experimentations under various agro-climatic conditions, the requirements of fertilizer nutrients are worked out for the different vegetable crops and their varieties. It is of utmost importance to know the procedure of calculating the exact quantity of manures and fertilizers. The high and low values may not bring desired results and the crop may suffer either due to excessive dose or low dose of fertilizers.

Equation

Calculate the quantity of urea, single super phosphate (SSP) and murate of potash (MOP) for 10 ha farm of tomato. If recommendation is 120:60:40 kg N, P_2O_5 and K_2O per hectare.

Solution

We know, Urea contains N=46 per cent, SSP contains P_2O_5=16 per cent and MOP contains K_2O=60 per cent

$$\text{Fertlizer required (kg/ha)} = \frac{\text{Rate of application of nutrient (kg/ha)}}{\text{Nutrient content in the fertilizer}} \times 100$$

In the given case

$$\text{Urea (kg/ha)} = \frac{120}{46} \times 100 = 260.9$$

$$\text{SSP (kg/ha)} = \frac{60}{16} \times 100 = 375.00$$

$$\text{MOP (kg/ha)} = \frac{40}{60} \times 100 = 66.6$$

For 10 ha:

Urea = 2069 kg

SSP = 3750 kg

MOP = 666 kg

Problem

Quantity of fertilizer per kg of nutrient

(a) For 1 kg of N   :

Calcium ammonium nitrate   : 4.0 kg

Amonium phosphate   : 5.0 kg

Urea   : 2.2 kg

(b) For 1 kg of P_2O_5 :

Single superphosphate   : 6.3 kg

Diammonium phosphate (DAP) : 2.2 kg

(c) For 1 kg of K_2O :

Murate of potash   : 1.7 kg

Sulphate of potash   : 2.1 kg

(d) 8.3 kg of (12 : 32 : 16) complex fertilizer would supply 1 kg N, 2.6 kg P_2O_5 and 1.3 kg K_2O

Conversion of Nutrient form

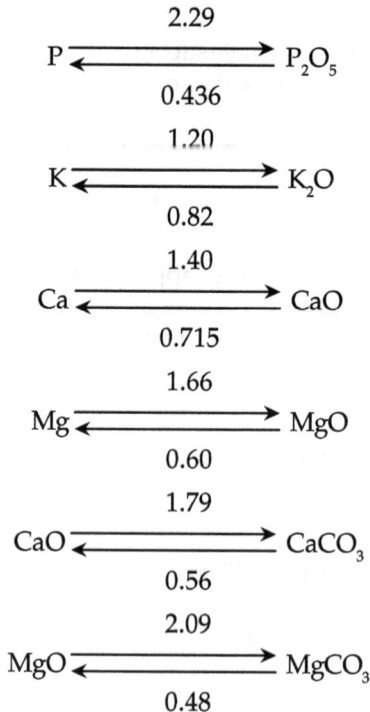

$$P \underset{0.436}{\overset{2.29}{\rightleftarrows}} P_2O_5$$

$$K \underset{0.82}{\overset{1.20}{\rightleftarrows}} K_2O$$

$$Ca \underset{0.715}{\overset{1.40}{\rightleftarrows}} CaO$$

$$Mg \underset{0.60}{\overset{1.66}{\rightleftarrows}} MgO$$

$$CaO \underset{0.56}{\overset{1.79}{\rightleftarrows}} CaCO_3$$

$$MgO \underset{0.48}{\overset{2.09}{\rightleftarrows}} MgCO_3$$

Chapter 9
Micro-Irrigation and Fertigation in Major Vegetable Crops

Introduction

Although water is a renewable resource, its availability in appropriate quality and quantity is under severe stress due to increasing demand from various sectors. Agriculture is the largest user of water which consumes more than 80 per cent of the country's exploitable water resources. The overall development of the agriculture sector is largely dependent on the judicious use of the available water resources. While the irrigation projects (major and minor) have contributed to the development of water resources, the conventional methods of water conveyance, and irrigation, being highly inefficient has led not only to wastage of water but also to several ecological problems like water logging, salinization and soil degradation making productive agricultural lands unproductive. Irrigation in green house is an essential part of successful vegetable cultivation. The hand application of water allows water application with a container having ¾th inch flexible hose @ 30 litre/minute but this practice is not very common due to high application rate which causes saturation of root zone area and more of disease incidence to the growing plants. The sprinkler method is very common for up land or low land areas where cultivation land is not leveled. However, sprinkler method is good for nursery raising and occasional above canopy. It has been recognized that use of modern irrigation methods like drip and sprinkler irrigation is the only alternative for efficient use of surface as well as ground water resources. Micro irrigation, which aims at increasing the area under efficient use of irrigation *viz.*, drip and sprinkler irrigation has therefore emerged at global level.

Micro irrigation is a system that supplies slow but frequent application of water in and around the root zone of the plant with the help of emitters/dippers. The term micro- irrigation is commonly used to describe several low-pressure irrigation systems, including drip/trickle, bubbler, micro-sprinklers and ultra-low rate irrigation system. Drip or trickle irrigation is most common and widely adopted system of micro-irrigation in country has increased steadily from 1500 ha in 1985 to 3 lakh ha in 1998. At present more than 75 per cent area under drip irrigation confines in the state of Maharashtra, Andhra Pradesh, Karnataka, Tamil Nadu and Gujarat. The potentially of drip irrigation in India is estimated to be 27 m ha. Vegetable constitutes one of the most important crops put under drip irrigation. Out of a total 1.8 lakh ha area under micro-irrigation (1991), field vegetables constitutes 8.9 per cent and green house vegetable comprises 3.6 per cent of total area. The leading country as regard to vegetable area under micro-irrigation is USA, Japan, Egypt, Israel, France etc.

The use of conventional irrigation methods not only results in considerable loss of water but also responsible for widespread salinity water logging, leaching of nutrient from rhizosphere and decline of water table. On the other hand, drip or trickle irrigation has proved its superiority over these methods owing to precise and direct application of water in the root zone without wetting entire area. Drip irrigation is most suitable tools for row planted wide spaced crops of high values especially in water scarce area.

Further fertilize is a costly input for crop production. Drip system of irrigation enables to apply the fertilizers along with water (fertigation) at a slow and controlled rate directly to the root zone. Fertigation gives better crop response, provides flexibility of fertilization, which enables the specific nutritional requirements of the crops to be met at different stages of crop growth.

Limitation of Micro-irrigation

In spite of technical and economic feasibility, its large scale and fast adoption is not so encouraging in India. The potentially of drip irrigation in India is of about 27 million hectare, the coverage so far is only about 3 lakh hectares. Fertigation yet to be adopted at a large scale due to non-availability of soluble fertilizers, apart from high cost of imported fertilizers. Drip irrigation systems are popular among horticultural crops confined to only in the west and South India (more than 80 per cent of coverage). In general, about 22 per cent of the area covered is under coconut. Other prominent crops are mango, banana, pomegranate, grapes, areca nut etc.

Fertigation

Fertigation is the application of water soluble solid fertilizer or liquid fertilizer through drip irrigation system. The factors that governs the fertigation are soil types, crops, methods of irrigation used, water quality, types of fertilizers available, economic feasibility etc. Fertigation has become an attractive method of fertilization in modern intensive agriculture systems. This has assumes added importance after the introduction of micro- irrigation system like drip in irrigated agriculture. Water and nutrient are the main factors of production in irrigated agriculture and are the major

Table 9.t: Irrigation Scheduling and Water Requirement in Vegetable Crops

Crop	Optimum Soil Moisture Regime	Total Water Requirement (mm)	No. of Irrigation	Irrigation Interval
Tomato	50-70 per cent available soil moisture (ASM) at 120 cm depth or at 0.7 to 1.2 PEC or at 65 kPa soil tension at 15 cm depth.	470-560	7-8	3-4 days in dry weather, 10–15 days during winter season
Brinjal	60-80 per cent ASM at 30 cm depth or 45 mm CPE	690-900	7-10	3-4 days in summer, 12-15 days in winter
Chilli and Capsicum	60 per cent depletion of ASM in 0-30 cm depth or at 65 kPa at 15 cm depth or 0.6 PEC	450-926	12-15	5-7 days in summer, 10-15 days in winter
Cabbage	50 per cent ASM in 0-30 cm depth or at 0.25 bar at 15 cm depth or at 1.2 PEC	400-450	8 (55 mm water each)	8-15 days
Cauliflower	50 per cent ASM at 30 cm depth or at 0.25 bar at 15 cm depth or 25-44 mm CPE	300-368	7	10 days
Radish and Carrot	0.4 PEC or- 20 k Pa at 18 cm depth or at 0.2-0.3 bar	200-250	3-8	3-5 days interval in summer, 10 days in winter
Onion	0.6-0.8 bar at 7.5 cm depth or IW/CPE=0.7 at 6 cm depth of irrigation water	550-725	12-15	5-7 days for main crop.
Okra	20-80 per cent ASM in 0-30 cm depth or 30-60 mm CPE	250-600	6-15	5-12 days
Cucurbits	50-75 per cent ASM or 0.9 PEC or -25 kPa at 15cm depth	400-500	9-10	7-10 days
Peas	50 per cent depletion of ASM	100-150	2	First at 45 days (flower initiation) and 2nd at 90-100 days (pod development)
Leafy vegetables	60-80 per cent ASM	250-350	5-7	4-5 days during summer, 7-8 days during winter

Table 9.2: The Comparative Advantages of Micro-irrigation over Conventional Method

Variables	Micro-irrigation	Conventional Method
Water saving	High :40-100 per cent	Less due to evaporation run- off, percolation etc
Irrigation efficiency	80-90 per cent	30-50 per cent
Input cost	Less in labour, fertilizer, pesticides and tilling.	Comparatively higher
Weed problem	Almost Nil	High
Water quality	Even saline water can be used	Only normal water can be used
Disease and pest problem	Relatively less	High
Water logging, run-off	Nil	High
Water control	High and easy	Less
Efficiency of fertilizer use	Very high and regulated supply	Heavy loss due to leaching
Range of applicability	In wide range of soil	Not suitable for sandy and undulated type of soil
Yield	20-100 per cent increase	Less compared to micro-irrigation.

Source: Biswas, B.C. and Kumar, Lalit FMN, 41(6): 3-4 (20¯0).

inputs in contributing higher productivity. In intensive agriculture, both fertilizer and irrigation management have contributed immensely in increasing the yield and quality of crops The method of fertilizer and irrigation application affects the efficiency of these inputs in arid and semi arid regions. Improvement of the use efficiency of these valued inputs is of utmost importance because these are costly and scare. Under this disadvantaged condition the use efficiency of these is also very low. Micro-irrigation systems are the most modern systems of irrigation where the use efficiency is very high and it is very popular in arid and semi-arid conditions of the world. Of late, it is also becoming popular in the arid and semi-arid region of India particularly where canal irrigation systems are not developed. With the advent of this new method of irrigation system, traditional method of fertilization, which is still in practiced by the farmers, is being slowly replaced by fertigation. In drip irrigation, the wetted soil volume and thus the active root zone is reduced under drippers and this small volume does not allow the addition of all plant nutrients needed by the plants. Rather, fertilizer needed is to be applied frequently and periodically in small amount with the each irrigation to ensure adequate supply of water and nutrient in the root zone. Therefore, because of the shift from surface irrigation to drip method of irrigation, fertigation becomes the most common fertilization in the irrigated agriculture. The use of soluble and compatible fertilizers, good quality irrigation water, and application of actual crop and water need are the prerequisite of the successful fertigation system. As any system has both advantages and disadvantages, so has the fertigation. The advantages and disadvantages of fertigation are mentioned below:

Advantages

1. In drip fertigation, fertilizer application is synchronized with plant need, which varies from plant to plant. In drip fertigation, the amount and form of nutrient supply is regulated as per the need of the critical stages of plant growth.
2. Saving in amount of fertilizer applied, due to better fertilizer use efficiency and reduction in leaching.
3. Optimization of nutrient balance in soils by supplying the nutrients directly to the effective root zones as per the requirement.
4. Reduction in labour and energy cost by making use of water distribution systems for nutrient application.
5. Better yield and quality of products obtained.
6. Timely application of small but precise amounts of fertilizers directly to the roots zone, this improves fertilizer use efficiency and reduces nutrient leaching below the root zone.
7. Ensures a uniform flow of water and nutrients.
8. Improves availability of nutrients and their uptake by crop.
9. Safer application method, as it eliminates the danger affecting roots due to higher dose.
10. Soil and water erosion are prevented.

Disadvantages

The main disadvantages are given below:

1. Both the components (drip and water-soluble fertilizer) are very costly.
2. Maintenance of drip irrigation is difficult. There is possibility of theft and rat infestation.
3. Good quality water is very essential. Clogging of emitters may cause a serious problem.
4. It needs water-soluble fertilizers; the availability of these types of fertilizers is limited.
5. Adjustment of fertilizers to suit the need is not easy.
6. Infestation of insects pest and diseases increases.
7. Area under micro irrigation is now increasing mainly because of subsidy in micro irrigation, if subsidy is withdrawn, the area under micronutrient may also reduced. So also would be the fate of fertigation.
8. Due to fear of yield loss, because of relatively lower dose of fertilizers in fertigation, farmers have the tendency to add additional fertilizers and irrigation water by traditional methods too. This may result in crop lodging (Sugar cane) lower yield and lower profits.

Components of Drip Irrigation

Drip irrigation system consists of following six components *viz.*,

1. **Power generating unit**: This unit is essential for supplying water from its source to the emission point.
2. **Filtration system**: To avoid physical clogging or blockage of emitters or drippers, filtration system is essential. Three types of fillers are used in drip irrigation system either used individually or in combination:
   (i) Sound filter to remove sand particles.
   (ii) Media filter to reduce both sand suspended materials.
   (iii) Screen filter to remove suspended particles effectively.
3. **Fertilizer–cum–agrochemical injection unit**: This unit consists of the fertilizer mixing tank which has inflow and antiflow pipes. It is connected to the main water line having a pressure regulation valve in between the two connecting points other agrochemicals like pesticides may be applied to the tank which is applied to the crop through irrigation water.
4. **Water distribution unit**: This unit is made up of network of a main line, submain and laterals. These lines or pipes are generally made up easier. High density polyethylene (HDPC) or low density polyethylene (LDPC), poly-vinyl chloride (PVC).

Sl.No.	Pipes	Diameter (mm)
1.	Main line	25-50
2.	Sub mains	15-35
3.	Laterals	10-20

5. **Water emitting unit:** This is made up of drippers (in-line drippers or online drippers) of 1 mm diameter in the laterals/lateral lines which are spaced according to spacing of the crop having a very low discharge rate *i.e.,* 1 to 4 litre/hour.

6. **Controlling unit:** This unit has the function of measuring, monitoring and regulating the flow of the water and other agro-chemicals running through the system which is ensured by means of simple valves to most sophisticated computerized control system.

Fertigation Equipments

Fertilizer can be injected into drip irrigation system by selecting appropriate equipment. Commonly used fertigation equipments are:

1. Venturi pumps
2. Fertilizer tank
3. Fertilizer injection pump

Venturi Injector

This is a very simple and low cost device. A partial vacuum is created in the system, which allows suction of the fertilizers into the irrigation system through venturi action. The vacuum is created by diverting a percentage of water flow from the main and passes it through a constriction, which increases the velocity of flow thus creating a drop in pressure. When the pressure drops the fertilizers solution is sucked into the venturi through a suction pipe from the tank and from there enters into irrigation stream. Although simple and with greater uniformity of dosing the fertilizers tank the venturi cause a high pressure loss in the system which may results in uneven water and fertilizer distribution in the field. The suction rate of venturi is 30-120 liter per hour.

Fertilizer Tank

In this systems part of irrigation water is diverted from the main line to flow through a tank containing the fertilizer in a fluid or soluble solid form, before returning to the main line, the pressure in the tank and the main line is the same but a slight drip in pressure is created between the off take and return pipes for the tank by means of a pressure reducing valve. This causes water from main line to flow through the tank causing dilution and flow of the diluted fertilizer into the irrigation stream. With this system, the concentration of the fertilizer entering the irrigation water charges continuously with the time, starting a high concentration. As a result, uniformity of fertilizer distribution can be a problem. Fertilizer tanks are available in 90, 120, 160 liters capacity.

Fertigation Pump

These are piston or diaphragm pumps, which are driven by the water pressure of the irrigation systems, and such as the injection rate is proportional to the flow of water in the system. A high degree of control over the fertilizer injection rate is possible, no serious head losses are incurred and operating cost is low. Another advantage is that if the flow of water stops, fertilizer injection also automatically stops. This is perfect equipment for accurate fertigation. A suction rate of pumps varies from 40 to 160 liter per hour.

Commercial Fertilizers Suitable for Fertigation

There are several commercial N, P and K fertilizers that can be used for fertigation. (Table 9.3).

Table 9.3: Nutrient Content of Common Fertilizers Suited for Fertigation

Nutrient	Compound	Nutrient Content in Solid Fertilizers $(N : P_2O_5 : K_2O)$	Nutrient Content in Saturated Liquid Fertilizers (25°C)
Nitrogen (N)	Urea	46-0-0	21-0-0
	Ammonium Nitrate	33-0-0	21-0-0
	Ammonium Sulphate	21-0-0	10-0-0
Phosphorus (P)	Phosphoric acid		
	Mono Ammonium Phosphate	12-61-0	4-18-0
	Diammonium Phospahate	18-46-0	

Source: Magor, H, *Fert. News*, 40(12): 97-100 (1995).

Crop Suitable for Fertigation

Fertigation can be practiced in large number of crops (Table 9.4). Row crops are most suited for fertigation.

Table 9.4: Crops Suited for the Drip Fertigation

Vegetable crops	Tomato, Chilly, Capsicum, Cabbage, Cauliflower, Onion, Okra, Brinjal, Bittergourd, Bottle gourd, Ridge gourd, Cucumber, Peas, Spianch, Pumpkin etc.
Spices	Turmeric, Cloves, Mint etc.
Plantation crops	Tea, Rubber, Coffee, Coconut etc.

Source: Rajput, T.B.S. Role of Water Management in Improving Agricultural Productivity. Indian J. Fertilizers 6(4), April (2010).

Area under micro irrigation is about 4.14 m ha (Table 9.5). However, what percentage of 4.14 m ha is employed for fertilization is not known.

Table 9.5: Area under Drip and Sprinkler Irrigation

Sl.No.	State	Drip	Sprinkler	Total
1.	Rajasthan	18455	731984	750439
2.	Maharashtra	505158	229590	734748
3.	Haryana	7904	529572	537476
4.	A.P.	400449	218066	618515
5.	Karnataka	190242	279942	470184
6.	Gujarat	182114	143598	325712
7.	Tamil Nadu	134378	27308	161686
8.	West Bengal	181	150171	150352
9.	M.P.	26117	125249	151366
10.	Chattisgarh	4175	72697	76872
11.	Odisha	4731	23844	28575
12.	U.P.	10778	10631	21409
13.	Punjab	12949	10677	23626
14.	Kerala	14585	2842	17427
15.	Sikkim	88	11033	11121
16.	Nagaland	0	4358	4358
17.	Goa	762	376	1138
18.	H.P.	127	639	766
19.	Ar. Pradesh	674	0	674
20.	Jharkhand	146	402	548
21.	Bihar	237	357	594
22.	Assam	116	129	245
23.	Mizoram	72	106	178
24.	Uttarakhand	38	6	44
25.	Manipur	30	0	30
26.	Others	16500	33000	49500
	Total	**1531007**	**2606574**	**4137581**

Source: Rajput, T.B.S. Role of Water Management in Improving Agricultural Productivity. Indian J. Fertilizers 6(4), April (2010).

Some Examples of Use of Drip Fertigation

Drip fertigation is found to be well suited for horticultural crops. In India, the adoption of this twin technology has resulted in enhancement of horticultural production. Water use efficiency and yield enhancement in vegetable production through drip irrigation are mentioned in Table 9.6.

Table 9.6: Water Economy and Yield Enhancement in Vegetable Production through Drip Irrigation

Vegetable Crops	Yield Increase (per cent)	Water Saving (per cent)
Tomato	50-60	40-60
Potato	20-30	40-50
Brinjal	20-30	40-60
Chilli	30-40	60-70
Cauliflower	60-80	30-40
Cabbage	30-40	50-60
Bottle gourd	30-40	40-50
French Bean	55-65	30-40
Okra	25-40	20-30

Source: Soman, P. Improving Water Use Efficiency to Enhance Crop Productivity. FAI Annual Seminar (2009).

Efficacy of Drip Fertigation

The benefit of drip irrigation mainly depends on the practice of fertigation because drip has a special feature, which is absent in other system of irrigations. In drip fertigation, the emitters moisten only 30-40 per cent of the soil. This is true in case of orchard crop. If the fertilizers and water are applied separately, the fertilizer use efficiency decreases because the fertilizer nutrients are not dissolved in the dry zones where the soil is not wetted. As a result, the benefit is not fully expressed. This is why, traditional fertilization is not appropriated, not convenient and efficient as the drip fertigation. Drip fertigation is therefore, the best means of fertilization to the root zone of the crops (Table 9.7). Use of drip fertigation is also very effective in high value row crop like sugarcane. Fertilizer use is quite high in sugarcane. In traditional method of fertilizer application large quantity of fertilizers are either banded near the crop row or broadcast in the inter-row space. Water is then allowed to flow through inter row space. Thus, a major portion of fertilizer is washed away to the end of the field or to the side of the drain. The localization of the nutrient at the root zone is very poor. Therefore, the fertilizer use efficiency is also very low. In addition, fertilizer application is limited to 2-3 splits applications during the life cycle of the crop limiting its availability to certain periods, which reduces the fertilizer absorption and use. Fertilizers need to be placed at a certain depth in the soils to be effective. The traditional method of application is inefficient because a high dose of fertilizer is placed on the surface and furrow irrigated. In the fertigation the nutrients are placed at very low concentrations through water move down into the lower soil layers where the absorptive roots exist.

Table 9.7: Fertilizer Use Efficiency in Fertigation (per cent)

Nutrient	Soil Application	Drip+Soil Application	Drip+Fertigation
N	30-50	65	95
P_2O_5	20	30	45
K_2O	60	60	80

Source: Soman, P. Improving Water Use Efficiency to Enhance Crop Productivity. FAI Annual Seminar (2009).

Layout of Drip Irrigation System

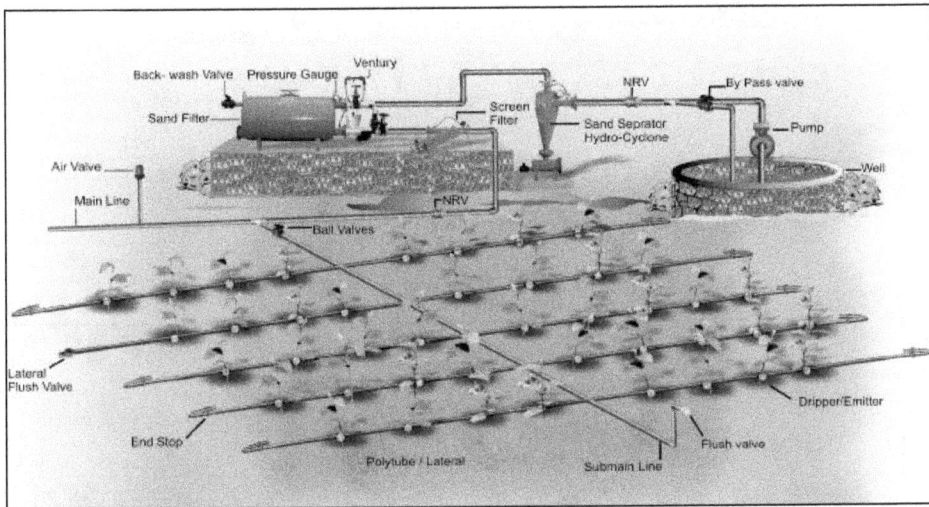

Sprinkler Irrigation

Under sprinkler irrigation, water is sprinkled under pressure into the air and plant foliage through a set of nozzles attached to network of aluminum or High Density Poly Ethylene (HDPE) pipes in the form of rainfall. These systems are suitable for irrigating crops where the plant density is very high where adoption of Drip Irrigation Systems may not be economical. Sprinkler irrigation is suitable for horticultural crops like vegetables and seed spices.

Conventionally, sprinkler irrigation has been widely in use for irrigating Cereals, Pulses, Oil Seeds and other field crops.

The sprinkler systems sets, unlike drip system, are moveable. Hence, one sprinkler set could cover more than one ha by shifting from one place to another. Assistance for sprinkler irrigation will be limited to only those crops for which drip irrigation is uneconomical.

Layout of Sprinkler Irrigation System

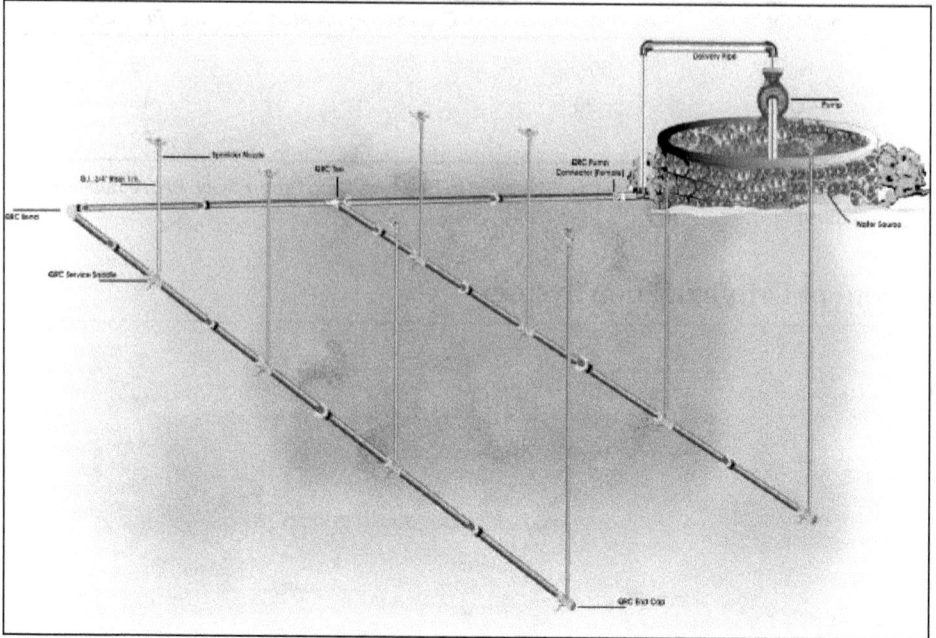

Chapter 10
Role of Mulching in Vegetable Crops

Mulching

Technically means covering of soil. Mulch provides the conditions that are favourable for the growth of plant and crop production. A material used to cover the soil to create a favorable microclimate among soil-water-plant is called mulch. We can say that any substance spread on the farm to protect the plants and crops is called mulch. Mulch can be natural mulches or synthetic mulches.

Definition

Mulching is the process or practice of covering the soil/ground to make more favourable conditions for plant growth, development and efficient crop production

(or)

Mulching is an agricultural cropping technique that involves placing organic or synthetic materials on the soil around plants to provide a more favourable environment for growth and production.

Advantages

- ☆ It prevents the direct evaporation of moisture from the soil and thus limits the water losses and conserves moisture. By evaporation suppression, it prevents the rise of water containing salts.

- ☆ Mulch can facilitate fertilizer placement and reduce the loss of plant nutrient through leaching.

☆ Mulches can also provide a barrier to soil pathogens.

☆ Opaque mulches prevent germination of annual weeds from receiving light

☆ Mulches maintain a warm temperature even during night time which enables seeds to germinate quickly and for young plants to rapidly establish a strong root growth system.

☆ Synthetic mulches play a major role in soil solarisation process.

☆ Mulches develop a microclimatic underside of the sheet, which is higher in carbon dioxide due to the higher level of microbial activity.

☆ Under mulch the soil structure is maintained during cropping period

☆ Early germination almost 2-3 days.

☆ Less nematodes population.

Types fo Mulches: Mainly two types of mulches
(1) Natural mulch and (2) Synthetic mulch

Natural Mulch	Synthetic Mulch
Whatever organic matter is readily available and transportable.	Black Plastic Mulch
Organic Mulches	Clear plastic Mulch
Compost, manure	Infrared transmitting (IRT) Mulch
Crop residues: straw, dry grass, clippings, leaves, and other left-over	Reflective Mulches
Wood products: sawdust, wood chips, Shavings	Degradable Mulches
Wastes from agriculture processing or from forestry	
Newspaper or cardboard	
Living/Cover crop	
Native plants	
Legume plants	
Grass: White clover, Red clover, Perennial ryegrass.	

I. Natural Mulches

A. Organic Mulches

Organic mulches are those mulches which are made out of natural substances. Organic mulches are providing nutrients and enrich the soil as decays. Agro-waste like paddy straw is fall in the category of organic mulch. Over the time organic mulches will decompose and becomes part of the soil and adds organic matter to soil, helping the soil to better retain water and nutrients, results in healthier plants. The organic mulches are temporary and will have to be replenished from time to time as it decomposes with time.

1. Straw

Straw from rice, wheat, barley and other crops is widely available after harvesting. Straw is inexpensive and normally sold in compressed Sales. It is then prepared and chopped in different sizes according to requirement. Straw mulch is light in weight and easy to apply and use. It is used as winter mulch around tree or shrub roots and as summer mulch in vegetable gardens. Nowadays, Paddy straw is used for mulching on fields, it provide better conditions for crop cultivation. Straw has some potential problems when used as mulch. Straw mulches are highly flammable, contains grain seeds that can germinate, it lowers the soil nitrogen supply as it decomposes and has to replenish annually. Dry paddy straw can be easily blown by wind. On the other hand, it is cheap and effectively suppresses weeds reduces soil water losses, conserves moisture, and insulates well. They are biodegradable and neutral in pH. Another important use for straw in summer is to prevent fruit contacting soil and becoming infected. Use straw around tomatoes, brinjal, capsicum, cucumber, melons, pumpkin etc. to prevent damage. It may also reduce slug damage.

2. Bark Mulches

Bark Mulches are prepared from the by-products of pine, hardwood logs or cypress. Bark mulches resist soil of compaction. Like light weight mulches, it will not blow away. These mulches are very attractive and are readily available. Bark mulches decompose slowly therefore will have not to be replenished from time to time as other mulches.

3. Wood Chips

Wood chips are made by reprocessing used timber and many different kinds of trees. Wood chip mulches may contain seeds from trees and other plants that can sprout and create weed problems, wood mulches should be properly aged or composted. Mulch that has not been aged form organic acids during the decomposition process and could kill or tender stem of young plants. Wood chip mulches have a high carbon: nitrogen ratio, it may temporarily reduce the supply of soil nitrogen for plant uptake during mulch decomposing. This loss can be compensating by adding nitrogen fertilizer to mulched plants/crop. It is practiced to replenished wood mulches every 2 to 3 years.

4. Sawdust

Sawdust is by-product of wood processing. Sawdust mulch is recommended for acid-loving plants like blueberries, rhododendrons etc. Sawdust is low in nitrogen, so it can robs nitrogen from the soil as it decomposes like wood chips. To compensate this loss nitrogen fertilizer may be needed. Because of compaction and decomposition Sawdust layers it is necessary to fluff up and replenished sawdust mulch each season.

5. Pine Straw

Pine Straw acidifies the soil like sawdust mulch and used for acid-loving plants. Pine needles decompose slowly.

6. Shredded Leaves

Leaves have been shredded before using as mulch; otherwise they form a mat and block water movement into the soil that blocks free water and oxygen movement into the soil. To avoid this leaves should allow undergoing partially decomposition before using them as mulch. Shredded leaves mulch contributing nitrogen and other nutrients to the soil while decomposing.

7. Strulch

It is made of mineralised straw and lasts about 3 times longer than ordinary straw. It is very light to carry but clingy enough to stay put on the ground. The bulk is airy which Prevents weeds seeds growing, but it keeps soil beneath moist. You get 95 per cent weed reduction from a 1 inch layer. It lasts 3 times longer than straw.

8. Leaf Mould

Leaf mould can be homemade. Make leaf mould by gathering fall leaves into polythene bags. Yellowed tree leaves are loaded with calcium but may be short on nitrogen. This is solved by adding alternate layers of wilted comfrey leaves. Moisten the bags, close, allow air in by perforating, and leave to rot. Leaf mould is available after 12 to 18 months. It can speed up the process by fine shredding. Resistant leathery leaves might then be used to make organic mulch - leaf mould and compost.

10. Lawn Cuttings

Nowadays lawn cuttings are being recycled directly onto the lawn by smart mulch mowers. Composting is another use for them. But it could easily have too much grass for composting. Whatever happens doesn't throw them away. Use as mulch on shrubs. Lawn cuttings are a little messy to remove after placement so use them under trees, shrubberies and hedges. Most bulbs and corms: Onion, daffodils, hyacinth, tulips, crocus; will push through easily. If you fork lawn cuttings into top soil don't forget to supplement the nitrogen content. Lawn cuttings are good for suppressing weeds for feeding heat insulating and retaining soil moisture. Scatter in a layer not thicker than 1 inch. Problems can arise with thick compact layers that exclude oxygen from the soil surface.

11. Farmyard Manure

These days, composted farmyard manure can be bought in nice 50L bags. In essence manure is a highly nutritious organic soil amendment. It is well applied to roses, and before planting potatoes, and nitrogen hungry greens – pumpkins, cucumber, cabbage, and marrow. It is well used as organic mulch in 'no dig' gardening. But it can equally well be mixed into the top few inches of soil. Raw farmyard manure includes: horse manure, cow manure, pig manure. Horse manure has a reputation for containing weed seeds. Pig manure rots down quickly. Cow manure is rather good because digestion is more complete. Composting increases the pH from acidic to more normal and stabilises the nutrient content. Farmyard manure is best applied in late winter or spring.

How Organic Mulching Works

- ☆ Soil conditioning and feeding
- ☆ Mulch is a surface covering
- ☆ Protects soil structure and reduces erosion
- ☆ Maintains winter warmth summer cool
- ☆ Protects against summer drought - improves drainage
- ☆ Suppresses weeds

Properties of Good Organic Mulch
- ☆ Doesn't get water saturated
- ☆ Drains freely
- ☆ Permits good aeration
- ☆ Free from contamination by weed seeds or fungal pathogens
- ☆ Easy to make, or obtainable at a reasonable price
- ☆ Nitrogen content not too low relative to carbon content
- ☆ Easy to apply

Problems with Organic Mulch
- ☆ Thick layers of matted organic matter can use up oxygen and prevent its diffusion in to the soil. Keep layers of grass cuttings to no more than 1 inch.
- ☆ Keep damp decaying mulch away from direct contact with plants. It could cause rot.
- ☆ Some organic matter breaks down to cause an acidic reaction in the soil; *e.g.* pine needles, some wood chip mulches. Don't use around lime loving plants.
- ☆ The reduction of heat transfer from soil to air due to insulating mulch could cause air frost. This effects exposed branches, dormant buds, or leaves. A good soaking beforehand cold weather could alleviate this.
- ☆ Mulches may provide a good environment for pests. For example, slugs sometimes cause serious losses to crops such as beans when mulched.
- ☆ Harmful insects, mice, rats, rabbits and snakes may also find thick mulches an attractive habitat.
- ☆ Many organic types of mulch have relatively high carbon content. This means that soil micro-organisms will take nitrogen from the soil to balance their ration. This causes temporary soil nitrogen depletion.
- ☆ The solution is to leave organic mulch on the surface where it is intended to be alternatively add nitrogen to the mulch. Comfrey leaves, comfrey tea, worm compost tea, garden compost tea, diluted urine, will all help to redress the balance.

B. Newspapers

Paper is organic and it rots down in the garden. It is good material for suppressing weeds and keeping the soil moist. A whole newspaper at least 8 sheets thick should be used with heavier material: compost, weed-tops, and straw, to hold it in place. Newspapers are well applied to vegetable plots because it can easily plant through. They are also popular for use around fruit and under trees. Gardeners also report using it successfully over surface planted potato tubers where it retains moisture for good growth. Weed suppression is the big advantage of blocking out the light. Spring is the best time to apply it.

C. Cardboard

Brown cardboard is effective at suppressing weeds because it excludes light. Obtained from shops as used packaging material it is the free organic mulch alternative to using plastic sheeting.

D. Living Mulch System (LM)

A living mulch system (LM) is a farming system that is used for sustainability of vegetable production and ecological conservation. The living mulch system represents an alternative to low-till/no-till practices used in agricultural systems. No-till practices are usually associated with high levels of crop residues left on the soil surface. Using a LM system has been shown to achieve agro-ecosystem stability in vegetable crop production. The LM system is a useful farming practice for improving soil quality, controlling weeds, preventing plant damage by pests and disease and increasing crop production. Some previous studies using living mulches have shown increased N availability, organic soil matter, soil structure, water infiltration (decrease water runoff), reduce soil surface temperature and water evaporation, and increase soil productivity.

Living mulch is an established cover crop and a living ground cover that is interplanted and grown either before or with an annual main crop throughout the growing season. Planting vegetable crops in tropical regions poses agricultural problems such as maintaining adequate water supply, pest and disease management, and weed suppression. It has been shown that winter rye planted as cover crop can be integrated in vegetable production systems along with herbicide treatments as a sustainable approach to improve weed management.

When to Use Cover Crops (CCs) in Vegetable Production Cycle?

Cover crops can be used in many applications in the vegetable production cycle. First, CCs can be used as a main crop during the primary growing season and as a rotation crop. Second, as a companion crop or living mulch, the cover crop is planted between the rows of the cash crop. Third, CCs can be used as a "catch" crop for nutrients, planted after harvest of the main crop to absorb of nutrients. Finally, as an off-season crop grown to protect the soil during periods when weather conditions prevent more valuable crop production.

Choosing a Good Quality of Living Mulch (LM)

There are four important characteristics for choosing a good quality of living mulch. Those characters are rapid plant establishment to prevent soil erosion and to control weeds; adaptability and persistence is needed to allow for entrance into the field; tolerance of drought and low-fertility soils and low maintenance budget associated with mowing, fertilizer application, and chemical stunting. Additionally, living mulch ground covers can offer a practical and economic management alternative for resource-poor farmers. Proper management of living mulches is crucial to their successful contribution to crop production.

Beneficial Effects of Living Mulch

The application of living mulch provides several beneficial effects for the biotic and a biotic feature in the ecosystem.

☆ Living mulches are able to prevent soil erosion reduce surface water pollution, add organic matter, improve soil quality and productivity and control weeds.

☆ Living mulches can increase nitrogen levels by 38-220 Kg N/ha from legume plants and 14-90 kg N/ha from non-legume plants.

☆ Living mulch can protect the soil from water erosion by reducing the raindrop impact, reduce soil bulk density, increase total soil porosity, water holding capacity and soil aggregation.

Negative Impacts of Living Mulch

☆ Living mulch has also been shown to have negative effects on the main crop similar to weedy species.

☆ Negative impacts of living mulches have been reported to be enhancing disease conditions and insect pest and compete for moisture and nutrients.

☆ In addition, found that crops grown in living mulch systems resulted in lower yields and later maturity than crops grown conventionally. The lower yield and later maturity of the crop plant was shown to occur because of shading, lower ground temperature and competition for plant nutrients by the living mulch plant.

Management of living mulches in agricultural crop systems is needed to minimize the negative impacts of the living mulch. Highly selective pre-or post-emergence herbicide to overcome the potential problems associated with the use of a living mulch, while not harming the vegetable crop.

II. Plastic Mulching for Crop Production

Introduction

Plastic mulches a boon to horticulture Synthetic mulching is an agricultural technique that involves placing synthetic materials on soil around plants to provide a more favorable environment for growth and production.

Types of Mulch Film

A wide range of plastic films based on different types of polymers have all been evaluated for mulching at various periods in the 1960s. LDPE, HDPE and flexible PVC have all been used and although there were some technical performance differences between them they were of minor nature. Owing to its greater permeability to long wave radiation which can increase the temperature around plants during the night times polyethylene is preferred. Today the vast majority of plastic mulch is based on LLDPE because it is more economic in use.

Basic Properties of Synthetic Mulch Film

1. Air proof so as not to permit any moisture vapour to escape.
2. Thermal proof for preservation of temperature and prevention of evaporation
3. Durable at least for one crop season

Important Parameters of Mulch

1. Thickness

Thickness (μ)	Recommended Crops
20-25	Annual-Short Duration Crops
40-50	Biennial-Medium
50-100	Perineal-Long Duration (> 12 months)

2. Width

This depends upon the inter row spacing.

3. Perforations

Depend upon situation.

Selection of Mulch Film

The selection of mulches depends upon the ecological situations and primary and secondary aspects of mulching

Type of Film	Purpose for which Used
Perforated mulch film	Rainy season
Thicker mulch film	Orchard and plantation
Thin transparent film	Soil solarisation
Transparent film	Weed control through solarisation
Black film	Weed control in cropped land
Black film	Sandy soil
Black film	Saline water use
White film	Summer cropped land
Silver colour film	Insect repellent
Thinner film	Early germination

Mulching Techniques for Vegetables/Close Space Crop

Laying of Mulch Film in Vegetable Crops

Thin film of 20 to 25 microns is used for mulching short duration crops like vegetables. Required length of film for one row of crop is taken and folded at every one metre or required spacing of the crop along the length of the film. Rounds holes are made at the centre of the film using a punch or a bigger diameter pipe and a hammer. Alternatively a heated pipe end could be used. In case the plant spacing is less than one metre required number of holes could be made as per the spacing of crop. For example if the plant spacing is 45 × 45 cm the folding could be done at every 45 cm along the length of film. The holes are punched on two spots of the face of the film. Alternatively the folding may be done at every 90 cm and four holes could be punched. In case of machine laying the punching of holes is done by the machine.

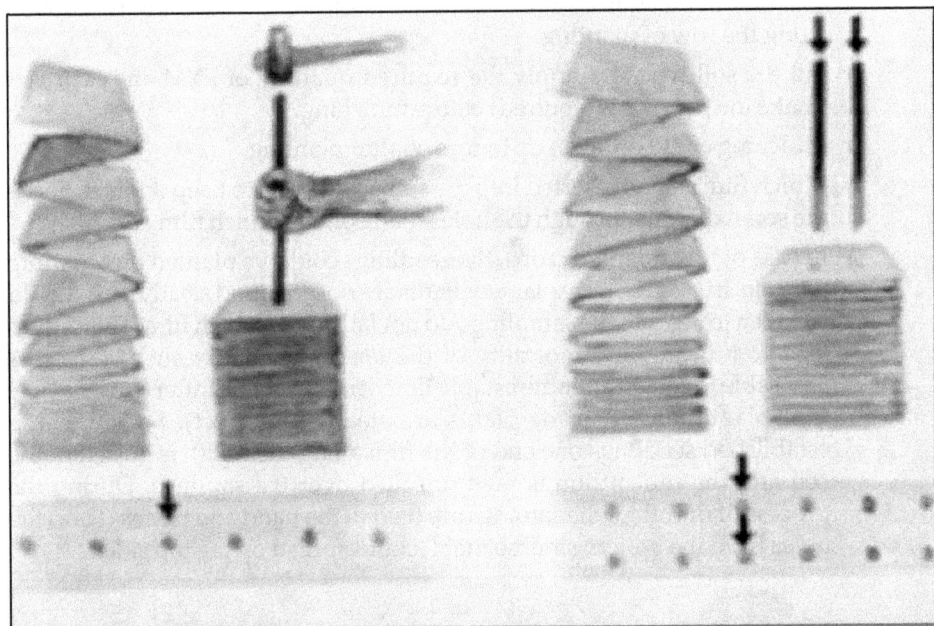

Punching of holes in films

Fixing of film on beds Planting/transplantation

☆ One end of the mulch film is anchored in the soil and the film is unrolled along the row of planting.

☆ Till the soil well and apply the required quantity of FYM and fertilizer (make the furrows is required) before mulching.

☆ Mulching could be taken up before or after planting.

☆ Mulch film is then inserted into the soil on all sides to keep it intact. Seeds are sown directly through the holes made on the mulch film.

☆ In case of transplanted crops, the seedlings could be planted directly into the hole. In case the transplanted plants are not erect and steady care should be taken to see that the samplings do not fall on the mulch film. This would lead to burning and mortality of the tender plants. In such cases it is advisable to mulch after the establishment of the plants (after 3 to 7 days of transplantation when the plants are steady and erect). For mulching established seedlings one end of the film along the width is buried in the soil and the mulch film is then unrolled over the saplings. During the process of unrolling the saplings are held in the hand and inserted into the holes from the bottom side, so that it could spread on the top side.

Mulching in vegetables

☆ In case the mulch film needs to be used for more than one season, the plant is cut at its base near the film and the film is removed and used.

Precautions for Mulch Laying

☆ Do not stretch the film very tightly. It should be loose enough to overcome the expansion and shrinkage conditions caused by temperature and the impacts of cultural operation.

☆ The slackness for black film should be more as the expansion; shrinkage phenomenon is maximum in this colour.

☆ The film should not be laid on the hottest time of the day, when the film will be in expanded condition.

Irrigation Practices under Mulching

☆ In drip irrigation the lateral pipelines are laid under the mulch film.

☆ In case inter-cultivation need to be carried out, it is better to keep the laterals and drippers on top of the mulch film and regulate the flow of water through a small pipe or through the holes made on the mulch film.

☆ In flooding the irrigation water passes through the semi circular holes on the mulch sheet.

Types of Plastic Mulches

(1) Black Plastic Mulch

Weeds do not grow beneath black plastic. Warm soil by 5 to 10 degrees. Beneficial in northern regions for warm season crops. Reduces soil compaction Ideal for drip irrigation.Black plastic mulches generally show very low reflectance and transmittance of short wave of short wave radiation but high transmittance of long wave radiation.

(2) Clear or Transparent Mulches

The transparent film will allow sunlight to pass through and the weeds will grow. However, by using herbicide coating on the inner side of film weed growth can be checked. The transparent film is quite successful as soil solarisation film for disinfecting the soil in order to reduce soil borne diseases and some weeds. This application is quite successful in nursery rising by solarising the beds before sowing seeds for nursery rising which gives 100 per cent seed germination and disease free

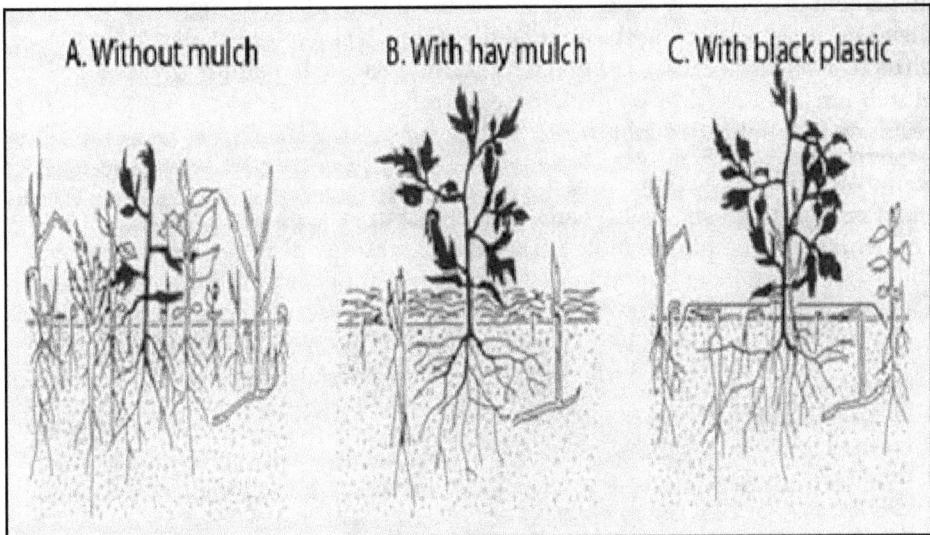

A. Without mulch B. With hay mulch C. With black plastic

nursery. While the black film has proved to be effective in plains to keep crop cool during summer, the transparent film is effective in hilly areas for raising soil temperature in cold climatic conditions during winter.

(3) Infrared Transmitting (IRT) Mulch

Infrared transmitting (IRT) mulch is a recent development. These plastics transmit the warming wavelengths of the sun, but not those that allow weeds to grow. These materials result in warmer soils than black plastic but cooler soils than clear plastics. The IRT mulches retard the growth of weeds including nut sedge. Crops grown on IRT mulch will develop 7 to 10 days earlier than crops grown on black plastic.

(4) Reflective Mulches

Insects developed tolerance to the insecticide consequently; alternative methods to protect vegetable crops from these insects are needed. Vegetable crops raised on beds covered with aluminium-painted mulch had a reduced incidence of insect transmitted virus infection, in comparison with crops raised on black or white polyethylene mulch, or on bare soil. For example, damage from mosaic virus transmitted by aphids (Aphididae) was reduced in watermelon and in squash. The incidence of tomato spotted wilt virus transmitted by thrips (Thripidae) was reduced 64 per cent in tomato and development of symptoms of To Mo V, transmitted by whitefly, was delayed on tomato. The reason for the reduced incidence of virus damage to the crops was that aluminium-painted mulch reflects light in the B (400- 500 nm) and in the near-ultraviolet (395 nm) regions of the spectrum. The light reflected in these regions repels the insects before they visit the plants.The aluminium surface mulch also emits less long-wave radiation to the leaf surface, resulting in 2 to 3 °C lower leaf temperatures than on black or on white mulch. The reflectivity of mulch might play an important role for energy and water conservation and might influence the plant growth. Reflective mulches alter crop environmental factors by increasing light intensity reducing soil temperature increase air temperature above the reflective surface. The reflective properties of a aluminum-painted plastic interfere with the movement of aphids by reflecting both short and long wave-lengths of light from their surface, disorienting the aphids' flight which spread the watermelon mosaic virus II. This virus causes the green streaking in squash. Painting the plastic with aluminum paint or white paint increases its reflectivity and cools late planted crops resulting in better fruit quality.

Removal of Plastic Mulch

Removal of plastic mulch namely discing, burning, physical removal, removal and storage of the plastic mulch. The plastic waste is disposed off through land filling, incineration, and recycling. Because of their persistence in the environment several communities are now more sensitive to the impact of discarded plastic on the environment, including deleterious effects on wildlife and on the aesthetic qualities of cities and forests. Improperly disposed plastic materials are a significant source of environmental pollution potentially harming life.

Photodegradable

Photodegradable plastic mulch an alternative solution for reducing waste from polyethylene mulches is to develop photodegradable or biodegradable. Photodegradable plastic mulch an alternative solution for reducing waste from polyethylene mulches is to develop photodegradable or biodegradable mulches. In the 1960s and 1970s, scientists started to investigate the possibility of using bio photo degradation as a self-destructive disposal technique for plastic film. Photodegradable mulches, which are manufactured with petroleum-based ingredients, to degrade into carbon dioxide and water have been questioned. Recently, newer photodegradable products have shown more satisfactory degradation characteristics when tested in different regions of the USA. Three major commercial products were:

☆ Plastigone an ultraviolet-activated, time controlled degradable plastic.

☆ Biolan an agricultural mulch film designed to photodegrade according to a predetermined schedule into harmless particles, which then biodegrade into carbon dioxide.

☆ Agplast a photodegradable material made by Lecofilms.

Photodegradable plastics are those reported to degrade by photo-initiated chemical reactions. The problem with these plastics is the continual use of non renewable petroleum based resources and their questionable ability to decompose to carbon dioxide (CO_2) and H_2O incompletely in the soil without light emission. Photo-biodegradable polyethylene films containing starch have been developed and used in agriculture. They are better able to raise temperature, preserve moisture, and raise yield than common polyethylene films and can be degraded environmentally after use. Starch-based mulch films have become popular in current research because starch is an inexpensive and abundant natural polymer that can produce a film structure.

The biodegradable mulches were easy to apply and were readily incorporated into the soil at the end of the growing season. Although the clear and to a lesser extent the wavelength selective forms of biodegradable mulch tended to break down well before the end of the growing season, this early failure did not negatively impact the performance of any of the crops tested, as long as supplemental weed control was provided. Although the biodegradable mulches are more expensive than the corresponding standard polyethylene-based plastics, this added cost is more than offset by the costs to remove and dispose of the standard plastic mulches.

Chapter 11
Important Vegetable Varieties/Hyrbids

Tomato

Open Pollinated Variety

Following varieties have been identified and recommended for cultivation in various zones:

Variety	Remark	Developing Centre	Recommended Zone	Year of Identification
S-12	Small fruited	PAU	–	1975
Pusa Ruby	Small fruited	IARI	–	1975
HS-101		HAU	–	1975
SL-120	Large fruited	IARI	–	1975
Sweet-72	Large fruited	Gwalior	–	1975
T-1	Large fruited	CSAUA&T	–	1975
Pusa Early Dwarf	Small fruited	IARI	–	1977
Sioux	Small fruited	–	–	1977
Sel-152	Small fruited	PAU	–	1977
Punjab Chhuhara	Small fruited	PAU	–	1977
KS-2	Det.	CSAUA&T	U.P.	1985

Variety	Remark	Developing Centre	Recommended Zone	Year of Identification
AC-238	Indet.	GBPUA&T	–	1985
CO-3	Det.	TNAU	I,IV,V,VI,VII	1987
Pb. Kesari	Det.	PAU	I,IV,V,VI	1987
La Bonita	Det.	NBPGR	I,IV,V,VII	1987
Pant T-3	Indet.	GBPUA&T	I,II,IV,V,VI,VII,VII	1987
Arka Vikas	Indet.	IIHR	I,IV,V,VI,VII,VIII	1987
Arka Saurabh	Indet.	IIHR	I,IV,V,VI,VII,VIII	1987
Sel-7	Det.	HAU	I,IV,V,VII,VIII	1990
Sel-1-6-4	Det.	PAU	I	1995
Sel- 32	Det.	HAU	II,VI,VII	1996
DT-10	Indet.	IARI	IV,VI	1996
BT-12	Indet.	OUAT	I,IV	1996
KS-17	Indet.	CSAUA&T	IV	1998
BT-116-3-2	Det.	OUAT	V, VI	2001
NDT-3	Det.	NDUA&T	IV, VII	2001
KS-118	Det.	CSAUA&T	IV	2001
DVRT-2	Det.	IIVR	VI	2001
BT-20-2-1	Indet.	OUAT	IV, VII	2001
NDT-9	Indet.	NDUA&T	IV	2001
NDTS-2001-3	Det	NDUA&T, Faizabad	IV	2004
Mani Laima	Det	ICAR Res. Com. for NEH Barapani	III	2004
IIVR Sel-1	Det.	IIVR	V, VII	2005
IIVR Sel-2	Indet	IIVR	IV	2005
BT-136	Indet	Bhubaneswar	II, IV	2005
VLT-32	Indet	VPKAS, Almora	IV	2005

Hybrid Variety

Following hybrids have been identified for cultivation:

Hybrid	Growth Habit	Developing Centre	Recommended Zone	Year of Identification
ARTH-4	Indet.	Ankur Seeds	IV,VIII	1992
MTH-6	Indet.	Mahyco	VII,VIII	1992
ARTH-3	Det.	Ankur Seeds	II,VII,VIII	1992
Pusa Hybrid-2	Det.	IARI	I,IV,VI,VII	1993
FMH-2 (A. Vardhan)	Indet.	IIHR	I, VII	1993
NA-501	Det.	Nath Seeds	IV,VII	1995
DTH-4	Det.	IARI	VII	1995

Hybrid	Growth Habit	Developing Centre	Recommended Zone	Year of Identification
KT-4	Indet.	IARI (Katrain)	IV	1995
NA-601	Indet.	Nath seeds	VI,VII	1995
FMH-2	Indet.	IIHR	IV	1995
Avinash-2	Det.	Novartis	VI	1998
HOE-303	Det.	Novartis	IV	1998
Sun-496	Indet.	Sungro	IV, II, VII	1999
BSS-20	Indet.	Beejo Sheetal	All	2001
DTH-8	Det.	IARI	IV	2001
CHTH-1	Det.	IARI	IV	2001
ARTH-128	Indet.	Ankur	VII	2001
KTH-2	Indet.	CSAUAT	IV, V	2003
JKTH-3055	Determinate	J.K. Seeds	I, IV	2004
KTH-1		CSAUAT	IV	2004
Nun-7730	Indeterminate	NunHems	I, IV	2004
TH01462	Determinate	Syngenta	I, II, IV, VI,VII	2005
HATH-5	Determinate	HARP, Ranchi	I	2008
ARTH-734	Indeterminate	Ankur	VIII	2008

Resistant Variety

Following disease resistant varieties of tomato have been recommended:

Variety	Resistant Against	Developing Centre	Recommended Zone	Year of Identification
BWR-5 (Arka Alok)	Bacterial wilt	IIHR	II	1992
FMH-1 (A. Vardhan)	Bacterial wilt	IIHR	V, VII	1993
FMH-2	Nematode	IIHR	IV	1995
BT-10	Bacterial wilt	OUAT	V,VI	1995
H-24	TLCV (Moderate)	PDVR	V	1997
BRH-2	Bacterial wilt	IIHR	VII	1998
LE-415	Bacterial wilt	KAU	I, V, VIII	2004
H86	TLCV	IIVR	I,IV,V,VIII	2005

Brinjal

Open Pollinated Variety

In brinjal, following open pollinated varieties have been identified based on their performance under various agro-climatic zones:

Variety	Fruiting Type	Developing Centre	Recommended Zone	Year of Identification
Pusa Purple Long	Long	IARI	IV,VI,VII,VIII	1975
Pusa Purple Cluster	Long	IARI	IV,V,VI,VII	1975
S-16	Long	PAU	IV,VI	1975
Pusa Kranti	Long	IARI	IV	1977
PB -129-5	Long	GBPUA&T	IV	1981
Pant Samrat	Long	GBPUA&T	IV,V,VI	1981
Arka Sheel	Long	IIHR	VIII	1981
Azad Kranti	Long	CSAUA&T	I,VI	1983
PB -91-2	Round	GBPUA&T	–	1985
ARU-1	Long	DARL	I	1985
T-3	Round	CSAUA&T	–	1975
H-4	Long	HAU	–	1987
Kat-4	Long	IARI, Katrain	VIII	1987
ARU-2C	Long	DARL	I,IV,VI,VIII	1987
K-202-9	Round	GAU	VI	1987
Aruna	Small Round	Akola	VII	1988
H-7	Long	HAU	IV,VI	1990
NDB-25	Long	NDUA&T	II, IV, VII	1990
H-8	Round	HAU	II, IV, V, VI	1990
BB-26	Long	OUAT	V	1993
Punjab Barsati	Long	PAU	IV	1993
Sel-4	Long	APAU	V	1995
DBSR-31	Long	IARI	VI	1995
KS-224	Round	CSAUA&T	I,II,IV	1995
DBR-8	Round	IARI	VI	1995
DBSR-44	Small Round	IARI	VI	1995
AB-1	Round	GAU	III,VI,VII	1996
PLR-1	Small Round	TNAU	IV,VI,VII	1996
BB-26	Long	OUAT	V,VII	1996
BB-13	Long	OUAT	VIII	1997
JB-64-1-2	Small Round	JNKVV	VII	1998
KS-331	Long	CSAUA&T	IV,V	1998
JB-15	Long	JNKVV	I	1998
CHBR-1	Round	CHES (Ranchi)	IV	1998
DBSR-91	Small Round	IARI	VII	1998
JB-64-1-2	Small Round	JNKVV	VII	1998
Green Long	Green	RAU (Sabour)	IV	1998

Variety	Fruiting Type	Developing Centre	Recommended Zone	Year of Identification
Punjab Sadabahar	Long	PAU	IV, VI	2001
NDB-28-2	Long	NDUA&T	IV	2001
DBL-21	Long	IARI, New Delhi	IV	2004
KS-235	Round	CSAUA&T,	IV, V and VII	2004
ABSR-2	Small Long	GAU, Anand	VII	2004
HABR-4	Round	HARP, Ranchi	IV	2005
IVBR-1	Round	IIVR	IV	2005
HABL-1	Long	HARP, Ranchi	I	2006
PB-66	Long	GBPUA&T, Pantnagar	VII, IV	2007

Hybrid Variety

In brinjal, following hybrid varieties have been identified, based on their performance under various agro-climatic zones:

Hybrid	Fruiting Type	Developing Centre	Recommended Zone	Year of Identification
Arka Kusumkar	Long	IIHR	VIII	1981
Arka Navneet	Round	IIHR	-	1981
Kat-4	Long	IARI (Katrain)	VIII	1987
Pusa Hybrid-6	Round	IARI	IV	1990
Pusa Hybrid-5	Long	IARI	IV, VII, VIII	1992
ARBH-201	Long	Ankur Seeds	IV, V, VI, VII	1993
NDBH-1	Round	NDUA&T	IV, VI, VII	1003
ABH-1	Small Round	GAU (Anand)	IV, VI, VII	1993
MHB-10	Small Round	Mahyco	IV, VI, VII	1993
MHB-39	Small Round	Mahyco	IV, VI, VII	1993
NDBH-6	Long	NDUA&T	IV	1995
ABH-2	Small Round	GAU	IV,VI	1995
ABH-2	Small Round	GAU	VII	1996
Phule Hybrid-2	Small Round	MPKV	VII	1997
Pusa Hybrid-9	Round	IARI	VI	1997
ARBH-541	Long	Ankur Seeds	All	2001
PBH-6	Long	Pandey Beej	All	2001
JBH-1	Round	Juna gadh	All	2001
BH-1	Round	Ludhiana	IV	2001
BH-2	Round	PAU, Ludhiana	IV, V	2002
VRBHR-1	Round	IIVR, Varanasi	IV, VI	2002

Hybrid	Fruiting Type	Developing Centre	Recommended Zone	Year of Identification
IVBHL-54	Long fruited	IIVR	IV	2004
ARBH-786	Long fruited	Ankur Seeds	IV	2004
VNR-51	Small round	VNR, Seed	IV, VI	2005
Navina	Long fruited	VNR Seeds	IV	2007
HABH-17	Round	Harp, Ranchi	IV	2007

Resistant Variety

Following bacterial wilt resistant varieties of brinjal have been identified for the cultivation under specific zones:

Variety	Developing Centre	Recommended Zone	Year of Identification
BB-7	OUAT	II, V	1990
BWR-12	IIHR	VIII	1990
SM-6-7	OUAT	VIII	1992
SM-6-6	KAU	I, VII, VIII	1996
BB-44	OUAT	V, VII	1996
CHES-309	HARP	I, VII	2001
BB-64	OUAT	IV, V, VII, VIII	2004

Chilli and Capsicum

Open Pollinated Variety

Following open pollinated varieties of chilli have been identified and recommended for cultivation under specific conditions and zones:

Chilli Variety	Developing Centre	Recommended Zone	Year of Identification
G-4	Lam	–	1975
G-5	Lam	–	1975
K-2	TNAU (K)*	–	1985
J-218	JNKVV	I, IV, V, VI, VII	1987
X-235 (LCA-235)	Lam	IV, V, VI, VIII	1987
Muslawadi	MPKV	V	1987
Sel-1	IIHR	V, VII, VIII	1990
LCA 206-B	Lam	V, VI, VII, VIII	1990
AKC-86-39	Akola	VII	2001
BC-14-2	OUA&T	V, VI	2001

Chilli Variety	Developing Centre	Recommended Zone	Year of Identification
RHRC-Cluster Erect	MPKV	VII	2001
PMR-57/88-K	IIHR	VII	2002
LCA334	LAM	III, IV, V, VII	2002
ASC-2000-02	GAU, Anand	VII	2004
KA-2	IIVR	IV	2005
LCA-353	RARS, Lam	V, VIII, IV	2007
BC-25	OUAT, Bhubaneshwar	VI, VII, V	2007

* TNAU - Kovilpatti

Hybrid Variety

Under hybrid trials following chilli and capsicum hybrids have been recommended:

Vegetable	Hybrid	Developing Centre	Recommended Zone	Year of Identification
Chilli	HOE-888	Sandoz	IV,VIII	1997
Chilli	ARCH-236	Ankur Seeds	IV	1997
Capsicum	KT-1	IARI (Katrain)	I	1990
Capsicum	SEL-II	Shalimar	I	1997
Capsicum	Lario	Syngenta	I	2001
Chilli	ARCH-228	Ankur Seeds	IV, V, VI	2002
Capsicum	DARL-202	DARL	I, IV	2002
Chilli	CCH-2	IIVR	II, IV,V,VI	2005
Sweet pepper	KTCPH-3	Katrain	I,VI,VII	2005

Garden Pea

Open Pollinated Variety

Following pea varieties have been identified for vegetable purpose:

Vegetable	Maturity Group	Developing Centre	Recommended Zone	Year of Identification
Bonneville	Mid Season	IARI	–	1975
GC-141	Mid Season	Gwalior	–	1975
GL-195	Mid Season	Gwalior	–	1975
Arkel	Early Season	IARI	–	1975
Early December	Early Season	Gwalior	–	1977
IP-3	Early Season	GBPUA&T	–	1985

Vegetable	Maturity Group	Developing Centre	Recommended Zone	Year of Identification
P-88	Early Season	PAU	–	1985
PM-2	Early Season	GBPUA&T	I	1987
Lincon	Early Season	IARI (Katrain)	I	1987
VL-3	Early Season	VPKAS	I,IV,VI	1987
VL-7	Early Season	VPKAS	IV	1992
Ageta-6	Early Season	PAU	I,IV,VI	1993
VL-6	Mid Season	VPKAS	IV	1993
PH-1	Mid Season	HAU	VII	1993
PH-1	Mid Season	HAU	V	1995
NDVP-8	Mid Season	NDUA&T	IV	1997
NDVP-10	Mid Season	NDUA&T	IV	1998
VL-8	Mid Season	VPKAS	I	1998
VRP-2	Early Season	IIVR	VI	2001
NDVP-12	Early Season	NDUA&T	IV	2001
VRP-3	Mid Season	IIVR	I	2001
Organ Sugar Podded	Edible Podded	PAU	VI	2001
VRP-5	Early	IIVR	I, IV, VIII	2005
CHP-2	Mid season	HARP, Ranchi	IV,VI	2005
VP-101	Early	VPKAS, Almora	IV, I	2007
PC-531	Mid season	PAU, Ludhiana	VI, VII, I	2007

Resistant Variety

Following powdery mildew resistant varieties have been identified:

Variety	Developing Centre	Recommended Zone	Year of Identification
PRS-4	CSAUA&T	IV,VI,VII	1990
JP-4	JNKVV	IV,VIII	1990
JP-83	JNKVV	VII	1992
NDVP-4	NDUA&T	IV	1995
DPP-68	Palampur	–	2001
KS-245	CSAUA&T	–	2001
NDVP-250	Faizabad	V	2002
DPP-9411	HPKV	I	2003
KTP-8	Katrain	I, IV, V	2005

Other Leguminous Vegetables

Open Pollinated Variety

Following varieties of legume vegetables have been identified for cultivation:

Vegetable	Variety	Developing Centre	Recommended Zone	Year of Identification
Cowpea	L-1552	–	–	1983
Cowpea	Sel-61-B	IHHR	VII,VIII	1992
Cowpea	Sel-263	PAU	IV	1992
Cowpea	Sel-2-1	NDUA&T	IV	1993
Cowpea	IIHR-6	IIHR	IV,VII	1998
Cowpea	NDCP-13	NDUAT	II, III, IV, VII	2002
Cowpea	IVRCP-1	IIVR	IV	2004
Cowpea	CHCP-2	HARP, Ranchi	VIII	2005
Cowpea	IIVR CP-4	IIVR	IV, V, VII	2007
Cowpea	VR-5	IIVR	IV, V, VII	2008
Cowpea	Swarna Harita	HARP, Ranchi	II, IV, V, VIII	2008
Dolichos bean	Deepaliwal	Akola	V,VII	1990
Dolichos bean	CHDB-1	HARP,	IV	2004
French Bean	VL-Boni-1	VPKAS	North Indian Hills	1985
French Bean	Arka Komal	IIHR	I,VII,VIII	1987
French Bean	UPF-191	GBPUA&T	IV,VIII	1987
French bean	IIHR-909	IIHR	I	1997
French Bean	CH-812	CHES	VII, III	2001
French Bean	OII-819	CHES	I	2001
French bean	IVFB-1	IIVR	I, VII	2005

Okra

Open Pollinated Variety

In okra, all the recommended open pollinated varieties so far have been identified under the disease resistance trial.

Hybrid Variety

Under hybrid trials following okra hybrids have been recommended:

Hybrid	Developing Centre	Recommended Zone	Year of Identification
DVR-1	IIVR	IV, VII	1998
DVR-2	IIVR	VI	1998
DVR-3	IIVR	All	2001

Hybrid	Developing Centre	Recommended Zone	Year of Identification
DVR-4	IIVR	IV, V, VII	2001
HBH-142	HAU	IV, V, VII, VII	2005
SOH-152	Syngenta	IV, VII, VII	2005
SOH-1016	Syngenta	IV,VII	2007
NBH-180	Nuzi Vedu Seeds	VII	2007
JNDOH 2-2	Junagadh	II, V, VI, VII, VIII	2008

Resistant Variety

YVMV resistant/tolerant okra varieties have been identified as follows:

Resistant/Tolerant Variety	Developing Centre	Recommended Zone	Year of Identification
P-7	PAU	All regions	1990
PB-57	Parbhani	All regions	1990
Sel-10 (A. Anamika)	IIHR	All regions	1990
Sel-4 (A. Abhay)	IIHR	II	1992
HRB-55	HAU	VI	1995
HRB-9-2	HAU	IV,VI	1996
VRO-3	IIVR		2001
VRO-4	IIVR	IV, V	2001
VRO-5	IIVR	VI	2002
VRO-6	IIVR	IV, V	2002
NDO-10	Faizabad	IV	2005
HRB-107-4	Hisar	VI,VIII	2005
IIVR-11	IIVR	VI,VII	2005
JNDOL-03-1	Junagarh	VII and VIII	2007

Onion

Open Pollinated Variety

Following varieties have been recommended for cultivation under specified zones:

Variety	Remark	Developing Centre	Recommended Zone	Year of Identification
Punjab Selection	Keeping quality	PAU	–	1975
Pusa Red	Keeping quality	IARI	–	1975
Pusa Ratnar	Red	IARI	–	1975

Variety	Remark	Developing Centre	Recommended Zone	Year of Identification
S-131	White	IARI	–	1977
N-257-9-1	White	MPKV	–	1985
N-2-4-1	Red	MPKV	–	1985
Line-102	Rabi Season	IARI	I, IV, VI, VII	1987
Arka Kalyan	Rabi Season	IIHR	IV,VI,VII,VIII	1987
Arka Niketan	Kharif season	IIHR	VII	1987
Agri Found Dark Red	Kharif season	NHRDF	IV	1987
VL-3	Rabi season	VPKAS	U.P. Hills	1990
Agri Found Light Red	Rabi season	NHRDF	VI, VIII	1993
Punjab Red Round	Rabi season	PAU	IV	1993
PBR-5		PAU	VI	1997
L-28	Rabi season	NHRDF, Nashik	IV and VII	2006
HOS-1	Rabi season	HAU, Hisar	VI	2006
B-780-5-2-2	Rabi season	NRC (O and G)	VI	2007

Garlic

Variety

Following varieties have been identified for cultivation:

Variety	Developing Centre	Recommended Zone	Year of Identification
G-41	NHRDF	IV,VII	1988
G-1	NHRDF	IV,VI,VII	1990
G-50	NHRDF	IV	1993
G-282	NHRDF	IV,VI,VII	1998
VLG-7	Almora	I	2001
DARL-52	DARL, Pithoragarh	I	2002
G-323	NHRDF, Nasik	VI	2002

Cauliflower

Open Pollinated Variety

Following open pollinated varieties have been identified for the cultivation:

Variety	Maturity Group	Developing Centre	Recommended Zone	Year of Identification
Early Kunwari	Early group	PAU	–	1975
327-14-8-3	Sep. maturity	IARI	–	1975
351-4-1	Oct. maturity	IARI	–	1975

Variety	Maturity Group	Developing Centre	Recommended Zone	Year of Identification
Improved Japanese	Nov. maturity	IARI	–	1975
EC 12012	Jan. maturity	IARI (Katrain)	–	1975
Pusa Snowball	Jan. maturity	IARI (Katrain)	–	1975
K-1	Jan. maturity	IARI (Katrain)	–	1979
114-S-1	–	GBPUA&T	–	1981
Line 6-1-2-1	Dec. maturity	IARI	–	1985
235-S	Nov. maturity	GBPUA&T	II, VIII	1990
KT-25	Snowball group	IARI (Katrain)	I	2001
IVREC-2	Early	IIVR	IV	2005
IVRMC-12	Mid-late	IIVR	IV	2008
DC-76	Mid-late	IARI, new Delhi	I and VI	2008

Hybrid Variety

Following hybrids have been identified for cultivation:

Hybrid	Maturity Group	Developing Centre	Recommended Zone	Year of Identification
Synthetic-1	Dec. maturity	IARI	–	1975
Early Synthetic	Early group	IARI	IV, VIII	1990
Pusa Hybrid-1	–	IIVR	II and IV	1992
DCH-541	Early group	IARI	II, IV	2003
SYCFH-202	Early group	Syngenta	IV, VII	2004
Summer King	Early group	Sungro	I, IV	2004
SYCFH-203	Early group	Syngenta	IV,V,VII	2005

Cabbage

Open Pollinated Variety

Variety Sel-8 has been recommended in 1985 for cultivation.

Hybrid Variety

Following hybrid varieties have been identified for cultivation:

Hybrid	Developing Centre	Recommended Zone	Year of Identification
Pusa Synthetic	IARI (Katrain)	IV, I, II	1992
Shri Ganesh Gol	Mahyco	V	1992
Nath-401	Nath Seeds	I, IV, V, VI, VII	1993
BSS-32	Beejo Sheetal	VII	1995

Hybrid	Developing Centre	Recommended Zone	Year of Identification
Nath-501	Nath Seeds	VII	1997
Quisto	Novartis	IV	1998
KGMR-1	Katrain	I, IV	2005
Green Emperor	Tokita Seeds	I	2007

Carrot

Open Pollinated Variety

During 1975, variety Sel-5 was recommended for cultivation. Other varieties identified are:

Variety	Developing Centre	Recommended Zone	Year of Identification
Sel-5	IARI	VII	1975
Pusa Meghali	IARI	VII	1992
SKAUC-50	SKUA and T, Srinagar	I	2006

Hybrid Variety

So far only one hybrid variety (Hybrid-1) developed at Mahyco, has been identified for its cultivation in zones I and VII.

Muskmelon

Open Pollinated Variety

Following varieties have been identified for their cultivation in specific agro-climatic zones:

Variety	Developing Centre	Recommended Zone	Year of Identification
Pusa Sarbati	IARI	IV	1975
Hara Madhu	PAU	IV,VII	1975
SI-45 (Pusa Madhuras)	IARI	IV,VI,VIII	1975
Arka Rajhans	IIHR	VIII	1975
Arka Jeet	IIHR	VIII	1975
Durgpura Madhu	Durgapura	IV,VI,VII,VIII	1975
NDM-15	NDUAT,Faizabad	IV	2002
IVMM-3	IIVR, Varanasi	IV	2006
GMM-3	AAU, Anand	IV and VII	2008

Hybrid Variety

Following hybrid varieties have been identified for cultivation:

Variety	Developing Centre	Source	Year of Identification
Hybrid M-3	IARI	IV	1990
MHY-5	Durgapura	VII	2001

Resistant Variety

Following disease resistant varieties of muskmelon have been identified:

Variety	Developing Centre	Source	Year of Identification
DMDR-1	IARI	CGMV	2001
DMDR-1	IARI	CGMV	2001
DMDR-2	IARI	Downy mildew + CGMV	2001

Watermelon

Open Pollinated Variety

Following varieties have been identified for cultivation:

Variety	Developing Centre	Recommended Zone	Year of Identification
Durgapura Meetha	Durgapura	IV,VI,VII,VIII	1975
Sugar Baby	IARI	V,VII,VIII	1975
Arka Manik	IIHR	IV,VII,VIII	1987
MHW-6	Mahyco	–	1999

Hybrid Variety

In 1981, one hybrid (Arka Jyoti) developed at IIHR has been recommended for cultivation.

Cucurbits

Open Pollinated Variety

Following open pollinated varieties have been identified in different cucurbitaceous vegetables:

Vegetable	Variety	Developing Centre	Recommended Zone	Year of Identification
Ash gourd	IVAG-90	IIVR,	IV and VIII	2006
Ash gourd	PAG-72	Pantnagar	VIII	2006
Ash gourd	Pusa Ujwal	IARI	VIII	2007
Bitter gourd	Priya	KAU	II,VII,VIII	1992
Bitter gourd	RHRBG-4-1	MPKV	IV,VII	1998

Vegetable	Variety	Developing Centre	Recommended Zone	Year of Identification
Bitter gourd	KBG-16	CSAUA&T	IV	1998
Bitter gourd	PBIG-1	GBPUA&T	IV	2001
Bottle gourd	Pusa Naveen	IARI	VII	1992
Bottle gourd	OBOG-61	GBPUA&T	IV,VI	2001
Bottle gourd	NDBG-104	Faizabad	IV	2002
Bottle gourd	NDBG-132	Faizabad	VI	2004
Bottle gourd	SEL P-6	IARI, New Delhi	IV,VII	2008
Cucumber	CHC-2	HARP	IV	2001
Cucumber	CH-20	HARP	IV	2001
Cucumber	PCUC-28	Pantnagar	I, VII, VIII	2001
Pumpkin	CM-14	KAU	IV,V,VII	1987
Pumpkin	Pusa Vishwas	IARI	I,VIII,IV,V	1987
Pumpkin	Arka Chandan	IIHR	VIII	1987
Pumpkin	Arka Suryamukhi	IIHR	VIII	1987
Pumpkin	CM-350	KAU	VII, VIII	2001
Pumpkin	NDPK-24	NDUA&T	IV,VI	2001
Ridge gourd	CHRG-1	HARP	IV	2001
Ridge gourd	PRG-7	GBPUA&T	VII	2001
Ridge gourd	IIHR-7	IIHR	VIII	2001
Sponge gourd	Sel-99	IARI	IV,VI	1995
Sponge gourd	CHSG-1	HARP	IV	2005
Sponge gourd	JSGL	Junagarh	VII	2005
Sponge gourd	KSG-14	CSAUA&T	IV	2006
Sponge gourd	PSG-40	Pantnagar	VII, I	2007

Hybrid variety

Following hybrid varieties have been identified in different cucurbitaceous vegetables:

Vegetable	Variety	Developing Centre	Recommended Zone	Year of Identification
Bitter gourd	Pusa Hybrid-2	IARI	IV, V, VI	2002
Bitter gourd	NBGH-167	Nirmal Seeds	IV	2004
Bitter gourd	Vivek	Sungro Seeds	VIII	2008
Bottle gourd	PBOG-2	GBPUA&T	VII	2001
Bottle gourd	PBOG-1	GBPUA&T	–	2001
Bottle gourd	NDBH-4	NDUA&T	All	2001
Cucumber	PCUCH-1	GBPUA&T	All	2001
Cucumber	Hybrid No. - 1	Century Seeds	I, IV, VII	2004
Cucumber	PCUCH-3	GBPUA&T	I, IV	2005

Chapter 12
Seed Production and Testing Procedure

Introduction

The primary objective of plant breeding is to develop better cultivars. The potential benefit of these cultivars is however, realized only if they are grown on a large area by the farmers. Therefore, it is essential that the seed of the improved cultivars is multiplied in sufficient quantity.

The seed production practices in the self and cross pollinated crops vary greatly on account of the amount of natural self and cross pollination. In self pollinated crops like pea, methi, tomato, lettuce, beans different cultivars can be grown side by side without much danger of contamination. On the other hand cultivars of cross pollinated crops have to be grown in isolation to avoid contamination. Secondly, the seed of standard cultivars/released cultivar is raised over and again, thus reasonable care is required to be exercised in maintaining the purity of cultivars. In case of the hybrids fresh seed has to be produced each year.

Objective

To produce the seed of the improved cultivars of pea, cucumber hybrid and onion cv. Agrifound Dark Red.

Assignment

- ☆ Record the percentage of off type plants in the pea cultivar (if any) also note down the plants of other crops, weeds and diseased plants.

☆ Perform the operations of removal of male flowers and bag female flowers of seed bearing parent and pollinate with the pollen from the male parent to produce cucumber hybrid seed.

☆ Do selfing and crossing in onion.

Procedure

Seed Production in Self-Pollinated Crops

The original seed of the cultivar, namely the breeder's seed is usually small in quantity and is under the control of the breeder. This seed is further multiplied at the farms of the University or by the State seed Corporations and is called the foundation seed. The foundation seed is further multiplied and is certified by the certification agency. Since the cultivars in the self pollinated crops are pure lines and the amount of natural cross pollination is very small. The cultivars in these crops should remain pure year after year. However, the crop should be kept free from off-type plants of other crops and seed borne diseases. Normally a small strip around the seed field measuring about 3-5 meters wide may be harvested for fruit crop rather than for seed. If the crop is planted close to another cultivar of the same crop. The rouging of off-type and weeds will ensure the maintenance of genetic and physical purity of the cultivar.

Hybrid Seed Production in Cucumber

Production of Inbred Lines

The purpose of producing inbreds is to fix desirable characters in homozygous condition. The inbreds are produced by continuous selfing and selection of desirable plants. The inbreds are maintained by allowing free pollination in isolation or by sib-mating. The isolation distance is 1000 meters.

Production of Single Cross

For this purpose inbred lines are grown in each of the isolated field for producing the hybrid seed. The ratio of the female line (from which male flowers are removed) to the male line is 4:2. The removal of the male flowers is done before pollen shedding in the field by inspecting the lines every day during the flowering period. The seed harvested from the male line shall be seed of the pollen parent, which can be used next year as male parent again. The selfed seed of the seed bearing parent/inbred line (female parent) is to be produced under isolation for producing hybrid seed next year.

Hybrid Seed Production in Onion

In onion, cytoplasmic genic male sterile line can be used for the production of hybrid seed. For commercial hybrid seed production, the male-sterile line is grown along with the pollinator line to produce the hybrid seed. Usually four rows of the female (male sterile line) and two rows of the male (restorer/pollinator line) are grown alternately in the field which is isolated from other fields of this crop by a distance of not less than 1000 meters. The hybrid seed is harvested from the female line. The rouging of off-type plants in the male and female rows, pollen shedding plants should be removed.

Seed production of the parental lines is a tedious job. The male sterile line is grown in isolation along with maintainer line keeping an isolation plots be grown for the maintenance and multiplication of the restorer and the maintainer lines. In the production of the parental seed stocks, strict purity standards should be maintained to provide genetic purity and identity.

Observations

Seed Production in Pea

Variety:

Isolation distance from other varieties:

Source of seed:

Morphological characters:

OBSERVATION SHEET

Observation	Sample 1	Sample 2	Sample 3
Number of off-type plants			
Number of plants of other crops			
Number of diseased plants			
Number of weed plants			

Hybrid Seed Production in Cucumber

☆ Name of the single cross/hybrid:

☆ Name of the female line:

☆ Name of the male line:

☆ Ratio of the female and male lines:

☆ Morphological features of female line:

☆ Morphological features of male line:

☆ Date of starting the removal of male flowers:

☆ Date of completing the removal of male flowers:

☆ Node at which the first female flower appears:

Hybrid Seed Production in Onion

☆ Name of the hybrid:

☆ Name of the female parent:

☆ Name of the male parent:

☆ Ratio of female and male parent:

☆ Isolation distance:

☆ Morphological traits of female parent:

☆ Morphological traits of male parent:

OBSERVATION SHEET

Observations	Female Parent	Male Parent
Total number of plants sampled		
Number of off-type plants and their percentage		
Number of diseased plants and their percentage		
Number of shedders and their percentage		

Key Questions

☆ Enlist the self, cross and often cross pollinating vegetable crops. What isolation distance do they require?

☆ What are the categories of seed? Define

☆ What are the field and seed standards? Enlist for different vegetable crops.

Seed is the basic input in agriculture. The potential of high yielding cultivars may not be realized if the seed is not of good quality. Good quality seed helps to perpetuate the traits of the improved cultivar. Therefore, it is important that the seed possesses the quality are genetical purity, seed health, weed seed, other crop seed, germination percentage, moisture content, seed weight etc. the seed will deteriorate in qulatiy if these factors are not taken care of.

Objectives

To obtain accurate and reproducible results regarding the purity composition, the moisture content, the percentage of weed seeds and the expected percentage of normal seedlings under favourable conditions.

Assignment

☆ Draw a sample from a seed lot of tomato to conduct purity analysis.

☆ Carry out the germination test to study the percentage of seeds which can give rise to normal seedlings.

☆ Determine the moisture content in the seeds with a moisture tester.

Procedure

Sampling

The basic objective in sampling a seed lot is to draw a portion which is true representative of the entire lot *i.e.* the same constituents are present as are present in the original seed lot and in the same proportion. Such a sample would be the basis for analysis to determine purity, germination, obnoxious weed seeds and other quality factors. The quantity of the seed tested in the laboratory is only a fraction of the seed sample submitted for testing. The fraction which is called the working sample is obtained by successively dividing the submitted sample by means of a sample divider. The sample divider mix the sample thoroughly as well as divides it into two approximately equal parts. The International Seed Testing Association (ISTA) has prescribed the minimum weights for submitted samples and working samples for various crops as given in the Table 12.1.

Table 12.1: Maximum Lot Size and Minimum Sample Weight for different Vegetables

Sl.No.	Vegetables	Maximum Weight of Lot (Kg)	Minimum Sample Weight	
			Submitted Sample (gm)	Working Sample for Purity (gm)
1.	Tomato	10,000	15	7
2.	Brinjal and Capsicum	10,000	150	15
3.	Chilli	10,000	150	15
4.	Cucumber	10,000	150	70
5.	Pumpkin	10,000	350	180
6.	Longmelon	10,000	150	70
7.	Cabbage	10,000	100	10
8.	Cauliflower and Broccoli	10,000	100	10
9.	Knol-khol and Kale	10,000	100	10
10.	Okra	20,000	1000	140
11.	Indian Spinach	20,000	2000	250
12.	Celery	10,000	25	1
13.	lettuce	10,000	30	3
14.	Spinach	10,000	250	25
15.	Carrot	10,000	30	3
16.	Beets	20,000	500	50
17.	Radish	10,000	300	30
18.	Turnip	10,000	70	7
19.	Onion	10,000	80	8
20.	Pea	20,000	1000	900
21.	Bean	10,000	1000	1000
22.	Chicory	10,000	150	5

Seed Purity Analysis

Seed purity refers to the physical purity. The separation of the seed sample into different constituents is carried out to determine the percentage of pure seed, inert matter, other crop seeds and weed seeds.

The working sample should be examined quickly in order to remove small inert fractions by sieving or chaff and dirt with blowing screens of different sizes. After the preliminary separation, if any, the seed should e examined on the clean surface of purity, work board and classified as under:

Pure Seed and Inert Matter

According to the International Rules, pure seed shall refer to and include all varieties of each species under consideration as stated by the sender or found after laboratory tests. In addition to this, special criteria have been fixed to differentiate between pure seed and inert matter. For example, the legume seeds usually break along the line of cleavage between the cotyledons. The cotyledons of split French bean, cowpea, soyabean are classed as inert axis regardless of whether the young root-shoot axis is present. For other legumes, if at least half the seed and radical are present, the structure is classed as pure seed.

All immature seeds should be regarded as pure seed if they can be identified as the kind under consideration.

Other Crop Seeds

other crop seeds include seeds of plants grown as crops. The classification for immature damaged, diseased and empty seed of other crops is the same as that provided for pure seed.

Weed Seeds

Weed seeds are the seeds, bulblets or tubers of plants recognized as weeds by law, official regulations or by general usage.

Inert Matter

It includes seed like structure from crop and weed plants that are one-half of the original size or less, badly injured and under developed seed like structures of weeds, glumes, stems and other plant parts plus sand, dirt and other related substances.

Weighing Components and Calculating Results

The weighing of the individual components should be done on an analytical balance upto four decimal places. The percentage by weight of each separation is determined by dividing the weight of the individual fraction by the total weight and multiplied by one hundred.

Seed Germination

Germination is defined as the emergence and development of a normal plant from the seed embryo under favourable conditions. Cabinet germination with controlled temperature and humidity is used for this purpose. Germination is judged on the basis of healthy seedlings with well developed roots and plumule. For germination, filter papers, towel papers or sand are most widely used substrata. Paper is easy to handle, versatile, non-toxic and cheap. Sand has an advantage of being inexpensive and reusable, but it should be sterilized before using again. Proper moisture (per cent) and temperature (mostly for 10°C-35°C) should be used in the germination test.

Germination test is always carried out with seed counted at random from the pure seed fraction to be tested in replicates of 50 or 100 seeds. The testing of 400 seeds is recommended for seed control and seed certification samples. However, only 200 seeds may be tested for service samples. The first and second counts are usually taken with paper tests whereas a single and final count is made with sand test. At the first and subsequent counts, the seedlings which fulfil normal seedling conditions are counted. All hard seeds, diseased and abnormal seedlings and the un-germinated seeds are left until the final count when their number is recorded. Diseased seeds and seedlings which may affect healthy seeds may be removed before the final count.

Moisture Test

The moisture content of seeds is one of the most important factors influencing their viability in storage. It is desirable to know the moisture content of the seed lots just after harvest or before storage as aid to seed trade. For this purpose, mechanical moisture testers such as the Universal Moisture Tester are used. Moisture can also be determined by using the oil distillation method.

Observations

Purity Analysis

Weight	Percentage
Pure seed	
Seeds of other crops	
Weed seeds	
Inert matteer	

Germination Percentage

Weight	Percentage
Normal seedlings	
Abnormal seedlings	
Hard seeds	
Un-germinated seeds	

Moisture Content
☆ Moisture percentage
☆ Weighing component and calculating results

Key Questions
☆ What is seed?
☆ How much sample required for seed analysis
☆ How much moisture (per cent) in seed is prescribed under ordinary packing and vapour proof packing?

Chapter 13
Vegetable Post harvest Losses

Food security has two aspects: Production and Post-Production. Both are of equal importance as only a well-managed post production system allows the consumers to have access to the food produced. Post production activities are an integral part of the food production system involving a series of operations from the producer through to the consumer. These activities are multidisciplinary in nature, involving harvesting, handling, storage and processing, equipment manufacture, as well as food marketing and other off farm operations. The concept of a post harvest implies that post harvest operations are considered as integrated functions, rather than as separate entities. A manual on post harvest losses has been prepared by FAO which, focuses on individual post harvest operations, their inter linkages and problems. FAO in collaboration with a number of international institutions actively conducting work in the area of post production recently established a mechanism of co-operation with the objectives of exchanging information and experiences on post harvest operations and dissemination of this information to other potential users. The special action programme for the Prevention of Food Losses (PFL) was launched by FAO, in order to advise member which could maximize their value and minimize post harvest losses. The programme has also assisted many governments in installing laboratories and in directing applied research toward upgrading the "post harvest practices" within their countries.

A post harvest loss in any change in the quantity or quality of a product after harvest that prevents or alters its intended use or decreases its nutritional value such losses after harvest are a major source of human food loss. Both quantitative and qualitative losses take place in vegetable crops between harvest and consumption. Need of the hour is to minimize these losses, and to achieve this one has (1) understand the biological and environmental factors involved in deterioration and (2) use those post harvest technology procedure that will slow down senescence and maintain the best possible quality. Qualitative losses include losses in (a) edibility (b) nutritional quality (c) caloric value and (d) consumer acceptability of products. Qualitative losses are much more difficult to assess than quantitative losses, especially because standards for quality and consumers purchasing power in developing countries are different from those in developed countries. For example, elimination of defects for a given commodity before marketing is much less rigorous in developing countries that in developing country. This however is not necessarily bad, because appearance quality is some what over emphasized in developing countries. A vegetable that is misshapen or has some blemishes may be a tasty and nutritious some that looks perfect.

Extent of Losses

Fresh vegetables are inherently perishable. During the process of distribution and marketing, substantial losses are incurred which range from a slight loss of quality to total spoilage. Post harvest losses may occur at any point in the marketing process, from the initial harvest trough assembly and distribution to the final consumer. The causes of losses are many physical damage during handling and transport, physiological decay, water loss or sometimes simply because there is a surplus in the market place and no buyer can be found. Losses are high in country, because of the inherent difficulty of collecting and transporting small quantities of produce from numerous small farms, and trying to collect these into a large enough quantity for efficient domestic marketing or for export. Postharvest losses of vegetables in country are so high, and the causes of these losses are so diverse, that a great deal of research and training is needed if prevention measures are to be improved. The need for improvement is shown by the fact that in developing countries where there is still a poor infrastructure and a lack of marketing facilities, postharvest losses of fresh produce range from 20 to 50 per cent.

Table 13.1: Estimated Losses by the US National Research Council

Name of the Vegetables	Loss (per cent)
Onions	16-35
Cauliflower	49
Tomato	5-50
Potato	5-40
Lettuce	62
Sweet potatoes	35-95
Carrots	44
Cabbage	37

Post harvest losses of horticultural perishables between the production and retail distribution sites are estimated to range from 2 per cent to 23 per cent, depending on the commodity, with an overall average of about 12 per cent. Estimates of post harvest losses in developing countries are two to three times those of the developed countries. Less post harvest losses can result in reduction in the area needed for production, conservation of natural resources, and protection of the environment. Strategies for post harvest loss prevention include use of genotypes that have a longer post harvest life, use of an integrated crop management system that results in good keeping quality of the commodity, and use of a proper post harvest handling system that maintains the quality and safety of products. Vegetables contain more than 80 per cent water and even 5 per cent loss in water cause many vegetables to appear wilted or shriveled. Physical losses are due to injury resulting from poor handling and packing, transportation and storage conditions. The injury of vegetables favours for fungi and bacteria, causing the rotting of microbial spoilage. The levels of post harvest losses in vegetables have been frequently estimated and global figures show that such losses vary between 0 and 60 per cent. Losses occur at different times during the production and post harvest cycle of the crop. A USDA study reported losses of 10.3 per cent of vegetables during transits unloading and related marketing. These figures don not take into account post harvest losses on the farm during sorting, grading, packaging, storage and marketing or even at the consumer end of the chain. In India, during transportation and marketing the post harvest losses of vegetables are estimated to be around 25-30 per cent. Post harvest quality of fresh vegetables generally depends upon the quality achieved at the time of harvesting.

Table 13.2: Post harvest Losses of Vegetables in India

Name of the Vegetables	Loss (per cent)
Cabbage and Cauliflower	7.08-25
Garlic	0.9-2.7
Onion	25-40
Brinjal	8.04-13.94
Carrot	5.9
Bottle gourd	7-10
Radish	3-5
Potato	30-40
Beans	7.45
Pointed gourd	19.81
Chilli	4-35
Beet root	10-15
Okra	5-9.01
Tomato	5-34.7

Harvesting Practices

Vegetables, which have high moisture content, generally have poor storage character. Moisture deficits in the soil have been shown to influence ascorbic acid contents in vegetables. Tomato, pepper, and other vegetables have shown increased vitamin-C levels in the presence of drip irrigation. Increasing the water availability through the drip line resulted in increased tomato fruit colour, size and acidity but decreased total soluble solids concentrations. The quality and condition of produce sent to market and its subsequent selling price are directly affected by the stage of maturity and harvesting methods. Consumers always prefer fresh, tender and disease and insect free vegetables of attractive appearance. Vegetables should be harvested as and when they attain maximum size however, the basic rules are (1) plan harvesting preferably at cooler part of they day *i.e.*, early morning or late afternoon and the produce should be shifted to shade as early as possible. (2) Do not harvest the wet produce, as wet produce will over heat if not well ventilated and leads to decay (3) harvesting during hot period raises field heat of the produce, which will cause wilting and shriveling. The principles, for harvesting is crucial to its subsequent storage, marketable life and quality. These may be defined in terms of either physiological maturity or horticultural maturity.

Table 13.3: Days from Pollination to Maturity

Name of the Vegetables	Days to Harvest Maturity
Pumpkin	65-70
Okra	4-6
Brinjal	25-40
Capsicum	30-35
Bitter gourd	12-13
Cucumber	6-7
Muskmelon	28-30
Water melon	35-60
Peas	30-35
Bottle gourd	12-15
Ridge gourd and Sponge gourd	5-6
Tomato (mature green)	35-45
Tomato (red ripe stage)	45-60
Round melon	7-8
Cowpea	10-15

The pre-harvest production practices have a tremendous effect on the post harvest shelf life. Soil nutrient and water management has great impact on post harvest storage life of vegetables. Much of the variation encountered in nutrient content in vegetables can be attributed to soil fertility. Vegetables contain high levels of nitrogen; have poor keeping quality than the same variety of crop with lower levels. Nitrogen deficiencies in the soil generally result in decreased protein in the harvested vegetable. Adequate levels of soil nitrogen can result in improved quality of vegetables. It has been found that decrease in ascorbic acid and dietry fiber of harvested cabbage was correlated with increase in nitrogen fertilization and higher head moisture content. High rates of phosphorous produced higher sugar content in tomatoes, as measured by soluble solids concentration. Potassium can also have significant impact upon crop quality. Optimal levels of potassium elicit favourable response in vitamin C content. Increasing levels of total and titrable acidity were obtained in tomatoes with increasing concentration of potassium.

Guiding principle is to harvest at a stage, which will allow it to be at its peak condition when it reaches the consumer, it should be at a maturity that allows the produce to develop an acceptable flavor or appearance. It should be at a size required by the market and should have an adequate shelf life.

Marketing Process

The problem of production has been replaced with the problem of marketing with scores of farmers finding it cheaper to dump their produce than to cart it to the market. Now the small farmers market most of the vegetables while cereals are mainly grown for self consumption. Marketing is one of the most important, yet misunderstood business activities and frequently means different things to different people. To the retailer in the agricultural sector for example it is selling along with other inputs to the farmer. To the farmer it is simply selling what he produces on his farm. However, what ever the circumstances, a well defined sequence of events has to take place to promote the product and to put it in the right place, at the right time and at the right price for a sale to be made. In the absence of a proper marketing infrastructure, the enormous production and its potential is marred by colossal wastage, very low level of processing and non-availability of post harvest infrastructure. A study by Technology Information Costing and Asseeement Council (TIFAC) of the department for Sciences and Technology published wastage for vegetables up to 20-30 percent at the post harvest stages due to poor storage, transportation and lack of infrastructure and the inadequacy of the marketing set up. A marketing strategy for vegetables is much more complex because of the perishability of fresh produce. These factors place limits on the time, the vegetable produce can be held the distance it can be moved. The law of supply and demand regulates the market price of vegetables due to which the price increases if the supply falls below market demand and decreases if supply exceeds the demand.

Table 13.4: Percentage Loss of Vegetables during Marketing

Name of the Vegetables	Whole sale	Retail	Total
Cabbage	18.65	2.19	15.84
Cauliflower	8.44	3.74	12.28
Tomato	12.07	5.78	17.85

The most important objective of marketing is to sell vegetables at the time and location that bring the highest possible return. The immediate aim of a marketing improvement programme is generally to generate a higher income for farmers through:

☆ Reduced post harvest losses by the introduction of better handling methods.

☆ Achievement of higher market price because of improved quality of produce.

☆ Proper training in marketing strategies.

Marketing cooperatives should be encouraged among producers of major commodities in important production areas. Such organizations are especially needed in developing countries because of the relatively small farm size. Advantages of marketing cooperatively include providing central accumulation points for the harvested commodity, purchasing harvesting and packing supplies and materials in quantity, providing for proper preparation for market and storage when needed, facilitating transportation to the markets, and acting as a common selling unit for the members, coordinating the marketing program and distribution profits equitably.

Alternative distribution systems, such as direct selling to the consumer (roadside stands, produce markets in cities, local farmers, markets in the country side etc.) should be encouraged. Production should be maintained as close to the major population centers as possible to minimize transportation costs.

Minimization of Losses

Vegetables are highly perishable living commodities that continue to respire and transpirate even after harvest. The modified atmosphere packaging is a term, which refers to the storage of fresh vegetables in plastic films, which restrict the transmission of respiratory gases. This results in the accumulation of carbon dioxide and depletion of oxygen around the produce which may increase their storage life.

Pre-cooling is a means of removing field heat with the aim to quickly slow down the respiration and transpiration rate. Several methods of pre-cooling are used commercially. The specific methods depend on the crop, the marketable of storage life required as well as the economics. Vegetables are living organisms. Their condition and marketable life are affected by temperature, humidity, composition of atmosphere which surrounds them, the level of damage that has been inflicted on them and the type and degree of infection with microorganisms. The vegetables deteriorate during storage through loss of moisture, loss of stored energy (carbohydrates) and other food components (vitamins), physical losses through pest and disease attack, loss in quality from physiological disorders and fiber development. The proper storage conditions must be maintained. Controlled atmosphere storage is one of the means of supplementing refrigeration to minimize deterioration. The controlled atmosphere storage requirements may vary within species or even with commodities, which may be held at same temperature and relative humidity. Most of the recommendations refer to the actual maximum storage life of the vegetable crop in the specified conditions, so that the crop is still in apposition to allow it to pass through normal marketing channels.

The vegetables are packed on site and transferred to a vehicle for direct transport to a pre-cooler or directly to market. Field containers for harvesting must be of a size that can be conveniently carried by the harvest worker while moving through the field. Plastic buckets or other containers are suitable for harvesting fruits that are more easily crushed, such as cabbage are sent to large scale packinghouses for selection, grading and packing. To achieve the maximum storage for a crop or to reduce losses during its marketable life, it is essential to keep it at the most appropriate temperature.

Table 13.5: Recommended Storage Conditions vs. Storage Life of Vegetables

Name of the Vegetables	Storage Conditions and Life
Tomato	**Mature green fruit**: 13.3°C and 90-95 per cent relative humidity 0 per cent CO_2 and 3-5 per cent O_2
	Red fruit: 7-14°C and 90-95 per cent relative humidity for breaker to light pink stage 0-3 per cent CO_2 with 3-5 per cent O_2
Cabbage	0°C and 95-100 per cent relative humidity for 90-180 days 2.5-5 per cent CO_2 and 2.5-5 per cent oxygen at 0°C
Pea	0°C and 90-95 per cent relative humidity for 7-10 days 5-7 per cent CO_2 with 5-10 per cent oxygen for 20 days
Cauliflower	0°C and 90-95 per cent relative humidity for 20-30 days
Chillies, Hot peppers	10°C and 90-95 per cent relative humidity for 14-21 days
Cucumber	10°C and 90-95 per cent relative humidity for 10-14 days 5-7 per cent CO_2 with 3-5 per cent oxygen
Okra, Lady finger	10°C and 90-95 per cent relative humidity for 7-10 days 0 per cent CO_2 and 3-5 per cent O_2
Bitter gourd	10°C and 90-95 per cent relative humidity for 10-14 days 0 per cent CO_2 and 3-5 per cent O_2
Pumpkin	12.2°C and 70-75 per cent relative humidity for 84-160 days
Water melon	10°C and 85-90 per cent relative humidity for 14-21 days
Capsicum, sweet pepper or Bell pepper	10°C and 90-95 per cent relative humidity for 12-18 days, 0-3 per cent CO_2 and 3-5 per cent oxygen
Brinjal	8-12°C and 90-95 per cent relative humidity for 1-2 weeks

A study of varietal difference in post harvest losses of tomato found interesting difference between varieties. This implies that plant breeding may be useful approach to loss prevention. Breeders should always keep in mind the storage life after harvest, as well as performance in the field.

Chapter 14
Regulatory Framework for Varietal Testing and Released in India

Introduction

There have been spectacular advances in world agriculture over the last three decades. These have been reflected in the greatly increased production of important crops and commodities. Given the importance of agriculture in the economies of developing countries, there is a need for a dynamic agricultural research system to sustain the gains and make further advances.

Sustaining the gains from the Green Revolution has been a major concern of agricultural policy makers, particularly for those formulating policy in developing countries. These countries have had to develop sound management practices of their own, or adopt models tested elsewhere. One set of policies have been adopted enthusiastically from North America and Europe, for variety testing, release and dissemination of seed to farmers. This system, in the form adopted by developing countries, is governed by a set of national laws, scientific guidelines, norms, and standard practices which together may be termed 'Regulatory Frameworks'. It is designed to provide:

Have standard and uniform testing and release procedures

Provide the regulations needed for varietal release

Determine the area of adaptation, the recommendation domain, of a new cultivar, and produce data on which to base extension recommendations.

There is no doubt that some regulation is necessary to make sure that only good, appropriate, new varieties are promoted by government, and that farmers get good quality seed, in sufficient quantity, when they need it. But over-regulation, or poor regulation, can block the release of varieties that could be beneficial to farmers, and can prolong unduly the whole process for successful varieties, from the early testing stage to reaching the farmers' fields. Once a variety has been released, its seed multiplication is subjected to rigorous seed certification standards, involving logistically complex field inspections and laboratory testing, before certified seed can be made available to farmers. In addition, the marketing of the seed of cultivars, and dissemination to farmers, is controlled by legislation and government policies. The regulatory framework aims to keep a large proportion of seed production in the hands of the public sector, so that the supply of seed–so important to farmers – is not left entirely to market forces that may be inefficient. In India, regulation began to develop at around the same time as the first improved cultivars emerged from CIMMYT and IRRI. It has evolved in response to changing circumstances, and, on the whole, has served the country well. However, after nearly thirty years of regulation, the system needs a radical review to remove some of the obstacles that prevent, or delay, providing low-resource farmers with improved seed suitable for their needs.

The Development of a Regulatory Framework in India

In 1963, the Government of India (GOI) established the National Seeds Corporation (NSC). It was given the primary responsibility for foundation seed production, and for the storage and supply of seed of released high-yielding cultivars of cereals, pulses, vegetables, fibre and fodder crops. With the growth of the seed industry, progressive seed growers and producers in the private sector came to be associated with NSC in the seed multiplication programmes. The certification standards were maintained by the NSC alone. However, it was felt that the large seed requirements of the Indian farming community could not be handled single handedly, and there was a need to have more seed producing and supplying organisations in the public and private sectors.

The GOI declared seeds as an essential commodity under the Essential Commodities Act 1955. In October 1964, the official varietal release system came into existence with the formation of the Central Variety Release Committee (CVRC) in the Indian Council of Agricultural Research (ICAR) at national level, and State Variety Release Committees (SVRCs) at state level. The Seeds Act1 was enacted by the Parliament in 1966 (NSC, 1992). The Seeds Rules framed under the Act were notified in 1968. A Central Seed Committee (CSC) was established, as provided in the Seeds Act, 1966, under the Ministry of Agriculture and Cooperation. The functions of the CVRC were taken over by the CSC in 1969. To ensure the quality of seeds on sale, the notification of the kinds/varieties, as envisaged in the Seeds Act, commenced in 1969. The first notification was issued as Standing Order (S.O.) 4045 on the 24.9.1969.

Some of the provisions under the Seeds Act particularly relevant to this critique are:

☆ The establishment of a Central Seed Committee (CSC);

☆ The establishment/designation of a central seed laboratory, and, at state level, a state seed laboratory;

☆ The setting up of procedures for the notification of kinds or varieties of seeds;

☆ Powers to set minimum standards for seed germination and purity, and marking or labelling requirements;

☆ Prohibition of the sale or supply of seed of any notified kind/variety unless it fulfils the requirements of the act;

☆ Authorization for states to establish a seed certification agency with powers to grant certificates to seed traders, and suppliers, who meet specified standards,

☆ The creation of Appeal Authorities to rule on decisions by a certification agency that are disputed.

Responsibilities set out in the regulatory framework are executed through two parallel systems, which to some extent overlap, and should be complementary: the central system and the system for states. Each has its own functions, structures, and set of relationships with other bodies. Both are empowered to release varieties, but only the central system can notify varieties. The framework has helped farmers to get genuine seed, and served as a mould for shaping the seed sector, which has turned into a well-knit industry.

The Central Testing, Identification and Varietal Release Procedures

Before a variety is released and reaches to the farmer, the All India Co-ordinated Crop Improvement Project identifies the variety for release in its workshop as per the established norms for testing for its value for cultivation.

Presently, All India Co-ordinated Crop Improvement (AICCPs) Projects have been created for almost for all the crops or groups of crops (Paroda, 1992). AICCIPs follow a three tier system of multi location evaluation spread over a minimum of three years involving the following stages:

First year (IET, Initial evaluation trial)

Second year- Advance varietal trial –I (AVT-I)

Third Year- Advance varietal trial –II (AVT-II)

Evaluation of Test Entries

☆ The constitution of varietal trial and their conduct varies from crop to crop (Gill, 1992). However, the initial evaluation trial includes the following entries:

☆ The entries to be nominated must have undergone critical evaluation/ screening in the station/station trials conducted by the sponsoring breeder. Secondly the entries to be nominated of three check varieties comprising of the following is used.

(i)   National check

(ii) Zonal check

(iii) Location check

(iv) Qualifying check

☆ All the trials are monitored by a team of scientists deputed by project co-coordinator to record the observations on the uniformity of crop stand, disease and insect pest resistance, bird damage etc. the main composition of the monitoring team is:

1.	Project Director/Project Co-ordinator/PI/Zonal Co-ordinator	Team leader
2.	Breeder	Member
3.	Agronomist	Member
4.	Pathologist/Entomologist	Member
5.	Scientist from any other specified discipline	Member

☆ The observations recorded according to guidelines on the data books for further supply to the co-ordinators.

☆ The data received at the co-coordinators' cell is critically examined for the inclusion in the Annual Report.

☆ Outstanding performance for yield of a variety by a margin of 10 per cent over the best performing check is promoted to AVT-I. However, norms for production vary from crop to crop.

Advance Varietal Trails (AVT-I)

Advance varietal trial is constituted by the entries promoted from IVT on the criteria specified above. Limited number of entries in AVT-I (not exceeding 16) is tested along with a minimum of three checks comprising of national check, zonal check and local check. All these entries are evaluated in a randomized block design with 3-4 replications at the different locations. The monitoring is done by the same team as given for IVT. Besides the agronomic and morphological observations, the additional data may be generated by the co-operators on disease and insect-pest resistance and quality. Again if a variety gives significant superior performance by a margin of 10 per cent over the best performing check in combination with other attributes is promoted to next stage AVT-II.

Advance Varietal Trial (AVT-II)

Same steps are followed as mentioned under AVT-I. However, the additional data to be generated at this stage. Response to agronomic variables such data of sowing, population densities and weedicides may be recorded. Data on diseases, insect-pests, quality parameters and abiotic stresses may also be recorded and discussed during the workshop. If the variety gives outstanding performance over the check (by a margin of 10 per cent) besides having all the favourable attributes, then the proposal for identification of a given variety is submitted by the concerned breeder on a variety identification proforma (Appendix) specified by the ICAR (Paroda, 1992).

Variety Identification System

The proposal containing all available data for the variety is considered by the variety identification committee constituted by the ICAR which meet during each AICRIP workshop.

Recommendations are made for country wide release or for a specific zone or states. Afterwards, the sponsoring agricultural university/research institute then submits the proposal for its release and notification to central subcommittee with the support of the Project Director/Coordinator (Figure 14.1).

Once the Central subcommittee accepts the proposal the variety/hybrid is released for the specific state or zone. The release proposal also ensures the availability of enough seed stock for seed multiplication on at least 10 ha.

State Varietal Identification System

The State Seed Sub-Committee (SSSCs) is constituted by Central Seed Committee and these SSSCs have been given responsibility to set up a State Seed Certification Agency (SSCA) aState Seed laboratory and to appoint a seed analyst and seed inspectors.

The SSCS is responsible for the release of a variety in its own state on the basis of data generated by State Agriculture University. The concerned breeder along with agronomist, pathologist, entomologist and biochemist generate sufficient data (usually more than three years). Secondly, sample variety must be evaluated in All Indian Co-ordinated Crop Improvement Project Trials. Thirdly, on farm trial data for a year or two are collected by extension agencies of State Department of Agriculture. After having all the above information, the State Agriculture University deliberates on the release proposal of a variety in a series of meetings before recommending to the SSC. Once approved by the SSSC for release in a state the variety is requested to be notified for seed production purpose by the Central Sub-committee.

Central Seed Committee and Central Sub-Committee on Crop Standards, Notification and Release of Varieties

The CSC is a statutory body constituted by the Ministry of Agriculture, Department of Agriculture and Cooperation, GOI to advise central government and the state level governments on matters related to the implementation of the Seeds Act 1966, and to carry out other functions assigned to it by, or under, the Seeds Act. The Core membership of the Committee includes: a Chairman to be nominated by the GOI, eight members nominated by the Central Government to represent interests that the Government deems appropriate, of whom at least two are representative seed growers, one person to be nominated by the Governments of each State. Typically, the CSC constituted by the Ministry of Agriculture, Department of Agriculture and Cooperation, vide its order of 29th October 1993, has the following members all who serve a two year term and are eligible for re-nomination. The Central Seed Committee may appoint sub-committees. It has appointed a Central Sub-Committee on Crop

	Deputy Comm iSsioner (QC) Department of Agriculture and Cooperation: Member Secretary, Co-convenor	
Coordinators/Directors of AICCIPs		Directors of State Seed Certification Agencies, or their representatives

Agri. Commissioner, GOI	Central Sub-Committee of the Department of Agriculture and Cooperation GOI Chair person: DDG (Crop Sciences) of ICAR	Directors of Departments of Agriculture of all states, or their representatives
Representatives of seed industry, NSC, State Seed Corporations, private seed companies		Progressive farmers

Representatives of ICAR, ICAR institutes, NGOs

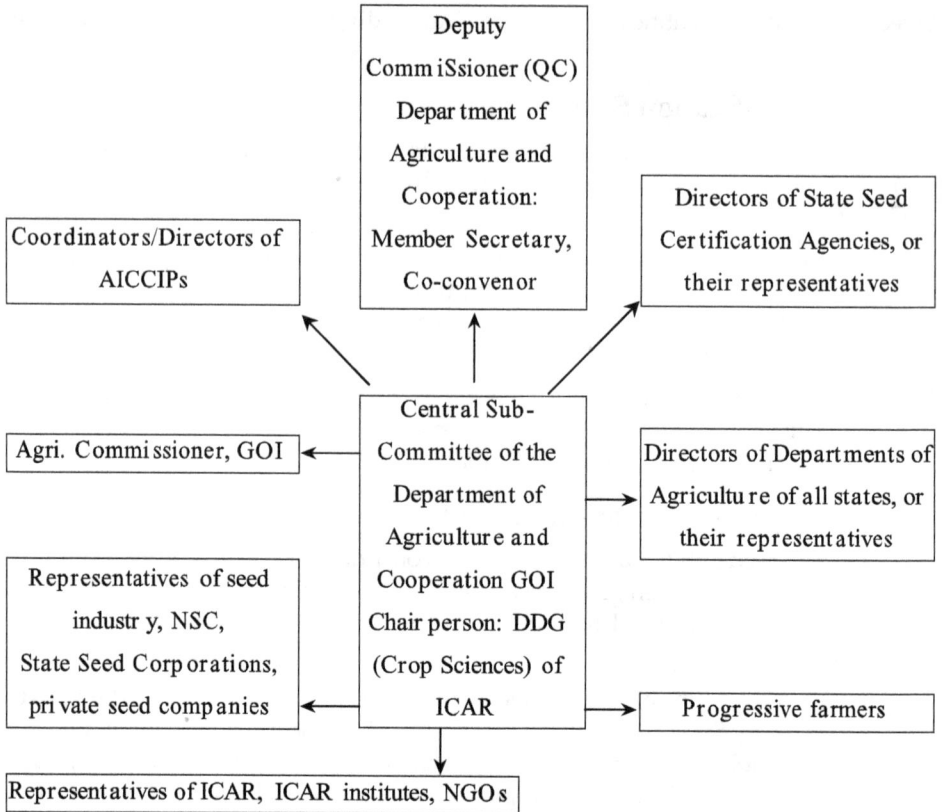

Figure 14.1: Constitution of the Central Sub-Committee on Crop Standards, Notification and Release of Varieties in 1994. The committee comprised a chairman and 17 members.

Standards, Notification and Release of Varieties. The composition of the Central Sub-Committee in 1994 (referred to as the Central Sub-Committee hereafter) is shown in Figure 14.1.

The terms of reference of the Central Sub-Committee are to:

☆ Approve the release of varieties/hybrids developed by the AICCIP, Central Research Institutes, the private sector, and other organisations provided the variety(ies) was/were considered suitable for more than one state.

☆ Approve proposals received from the State Variety Release Committees/State Seed Sub-Committees for varieties developed by the State Research Institutes but are considered suitable for areas outside the state.

☆ Delimit the regions or tracts for cultivation of varieties approved for release.

☆ Advise the ICAR regarding the manner in which the National Register of Approved Varieties may be maintained, and to suggest the standard description of varieties.

☆ Notify kinds/varieties for the purpose of the Seeds Act and the areas of their notification.

☆ Specify minimum limits of germination and purity for the notified kinds/varieties of seeds.

☆ Specify the "mark" or "label" in respect of notified kinds/varieties.

The Central Sub-Committee releases varieties of regional or national importance, and the State Seed Sub-Committees release varieties deemed to be beneficial for individual states. Release is intended to make a newly developed variety available to the public for general cultivation in the regions for which it is adapted.

The Central Sub-Committee is authorised by the Central Seed Committee to notify, by notification in the official gazette, kinds/varieties for the purpose of the Seeds Act and the areas for which these kinds/varieties are to be so notified. Varieties released by the State Seed Sub-Committees are also centrally notified after formal application to the Central Sub-Committee. Notification is a means to enforce the provisions of the Seeds Act on regulating the quality of seed. No certified seed can be produced by any seed multiplication agency unless the variety is notified. Notification usually authorises certified seed production throughout the country, by private or public seed multiplication organization.The cooperating centres of the coordinated projects contribute their most promising entries to the all-India coordinated trials. The entries in the first year are evaluated in the Initial Evaluation Trials (IET). In some crops, because of the large number of entries the IET is preceded by testing for one year in the National Screening Nursery. Based on the promotion criteria, the materials possessing higher yield (usually the major criterion) or better disease/pest resistance, drought tolerance and or quality traits are tested in the Advanced Varietal Trials (AVT). Along with testing in the AVTs, data are normally generated regarding agronomic responses, disease and pest reaction under hot spot and artificial epiphytotic conditions at selected centres (Figure 14.2).

Identification and Release

Release and notification of a variety follows its identification and recommendation by the AICCIP workshop after a minimum of three years of multilocational trials and assessment for Value for Cultivation and Use (VCU). The variety should be suitable for specified agro-climatic and soil conditions, have an ability to withstand typical stress conditions, and have tolerance/resistance to pests and diseases. It should also show distinct advantages over the existing equivalent released varieties.Trials data on agronomic performance need to be provided. For a proper assessment, data on performance against popular varieties on farmers' fields are also needed, but the degree to which such data are collected and included in release proposals varies; provision is not mandatory.

At the end of the AVT II stage, the proposal for identification of a variety is submitted by the concerned breeder on a variety identification proforma specified by the ICAR (Paroda, 1992). The form was standardised in 1992 (Annex 2), and are similar for central and state releases. Prior to this, there was a wide variation in

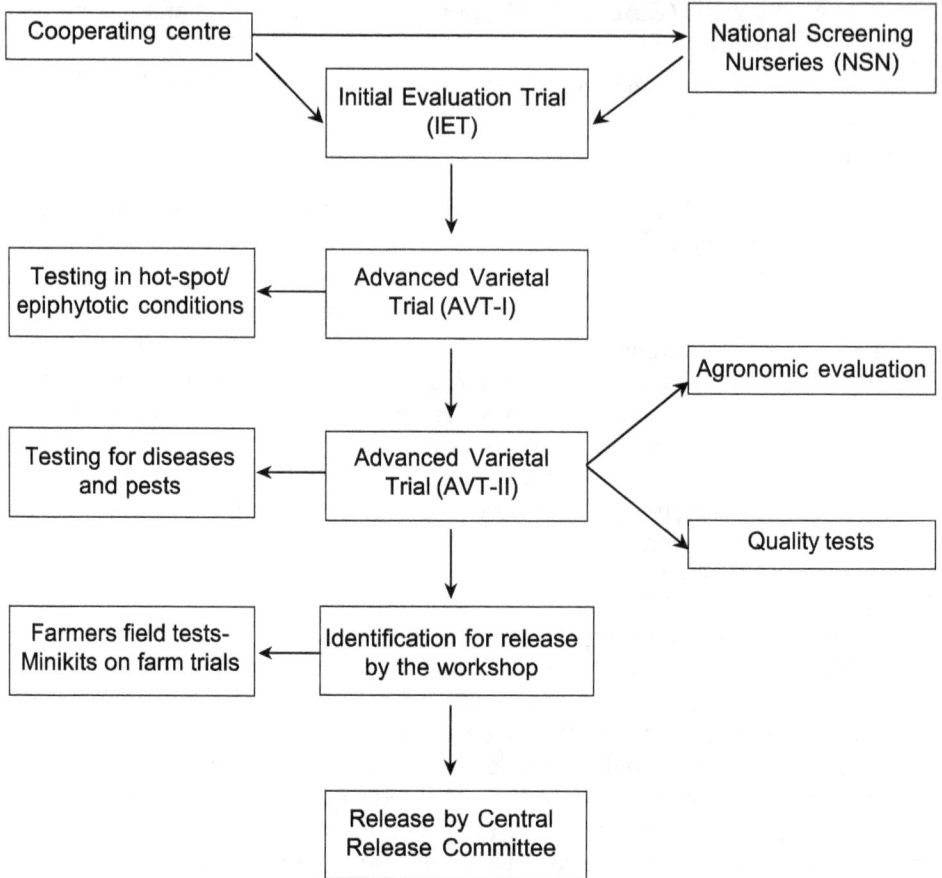

Figure 14.2: Procedure for Varietal Testing and Release in the AICCIPs.

presentation of data, and many important comparisons were not available between the proposed variety and the check cultivar. The proposal containing all available data for the variety is considered by the Variety Identification Committee constituted by the ICAR which meets during each AICCIP workshop (Figure 14.3).

The Principal Investigators and Zonal Co-ordinators attend the meeting to provide wider information on the variety. The Director of Crop Development Programme is invited to provide information on the response of farmers to minikit trials if they were conducted at the same time as the AVT II.

Recommendations are made for country-wide release, or for a specific zone, or states. Conditions for cultivating the variety are also specified, for example, late, early, or timely sowing; irrigated or rainfed cultivation; or high or low fertility conditions. Warnings may qualify the recommendations, such as details on susceptibility to diseases. The sponsoring agricultural university/research institute then submits the proposal for its release and notification to the Central Sub-

Committee, with the support of the Project Director/Coordinator. The proportion of varieties accepted by the Variety Identification Committee and those subsequently released by the Central Sub-Committee varies greatly across crops.

If the Central Sub-Committee accepts the proposal, the variety/hybrid is released for the concerned states (*i.e.*, for more than one state, and often country-wide) it is simultaneously notified for certified seed production purposes, usually for the entire country. However, a state for which it is released usually requires the variety to be tested in Adaptive Trials within the state before it can be formally recommended.

The release proposal proforma requires the breeder to ensure availability of enough seed stock for seed multiplication on at least 10 ha. Post-release seed multiplication is the responsibility of various seed agencies.

Seed is made available for minikit trials to help popularise the variety and evaluate the response of farmers. Minikit trials are organised by the Directorates of Crop Development, Ministry of Agriculture and Cooperation, GOI, and are conducted by the State Departments of Agriculture, but are funded by the GOI. Minikit demonstrations and trials are conducted with both released and pre-release varieties at the AVT II trials stage. In the latter case, this cuts overall testing time and can provide valuable information on how the variety performs on farmers' fields prior to its identification by the Variety Identification Committee.

A similar procedure applies to varieties produced by the private sector if they are intended to be officially released. However, it is not mandatory that a variety developed by the private sector be released centrally or by state committees, and private sector participation in the AICCIP trials is optional.

The seed certification rules are uniformly applicable to the public and private sector. The private sector takes the advantage of selling 'Truthfully Labelled' seed of any variety. Truthfully labelled seed is not field certified to assure genetic purity but the seed standards are not lower than the certified class of seed. Unreleased varieties (private sector or public sector) do not come under the purview of seeds act for the purpose of certification. For varieties of foreign origin, often imported by the private sector, the Seeds Rule of 1988 permits them to be provisionally notified after being tested for one season over 15-20 locations. For regular notification, an additional of two years testing will be required. For varieties of vegetable crops, flowers and ornamental plants there is an Open General Licence (OGL) where import and sale of seeds do not require any evaluation, release or notification.

The State Varietal Identification and Release System

A state research institute or a private seed company can attempt to release a variety through the central system, or the state system. If the variety is widely adapted, and suited to conditions beyond a single state, the breeder may try for central release; in other cases state release may be easier, and more appropriate.

The identification and release of varieties is regulated by national Acts, Rules and Orders, but there is considerable variation in implementation and practice between states. Some states assume considerable independence of action; others are more tied into the central system. Some test new varieties on farmers' fields, and seek

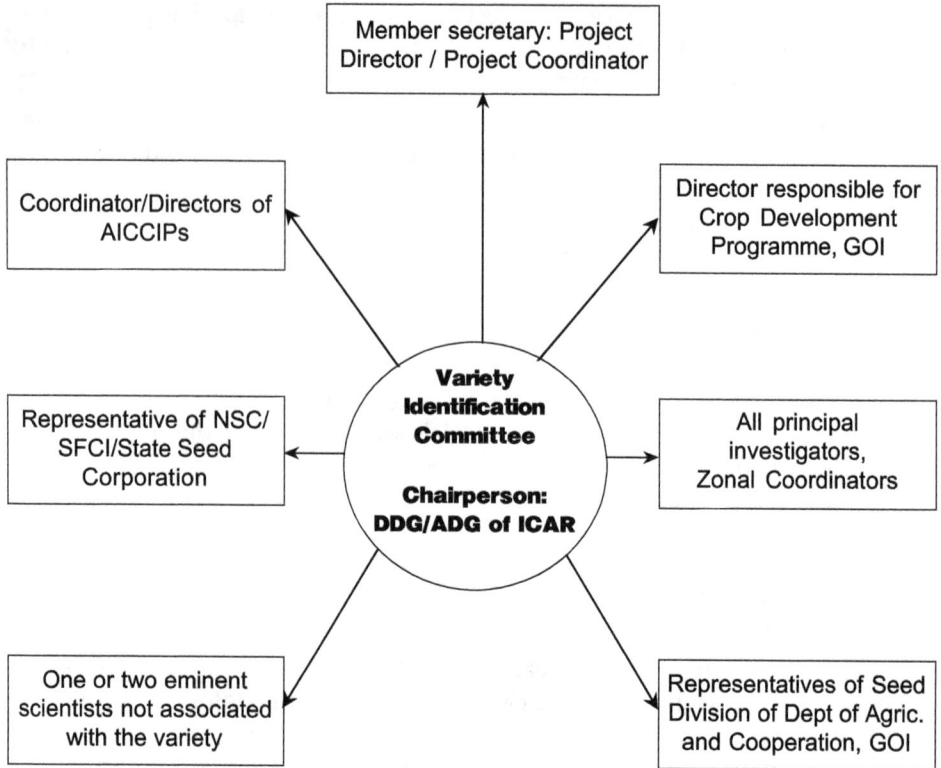

Figure 14.3: Constitution of Variety Identification Committee during an AICCIP Workshop.

the reactions of farmers early on; others make little or no attempt to involve farmers until after a variety is released.

Identification and testing of new varieties is usually mainly the responsibility of one or two agricultural universities. Popularization is the function of the Department of Agriculture and the agricultural universities, acting in collaboration and separately. Seed production, certification and distribution is undertaken by various state seed agencies, the university/universities, and private seed companies. To date, NGOs have played only a small part in the whole process. Overall authority lies with the State Seed Sub-Committee.

State Seed Sub-Committees

State Seed Sub-Committees (SSSCs) are constituted by CSC, and are delegated to set up a State Seed Certification Agency (SSCA), a State Seed Laboratory and an Appeals Authority, and to appoint seed analysts and seed inspectors. Typically membership of SSSCs is given in the subsequent three chapters. It is the duty of the SSSC to:

☆ Advise the state government on all matters relating to the implementation of the Seeds Act;

☆ Review the implementation of the Seeds Act in the state and send periodic reports to the state government and the Central Seed Committee;

☆ Inspect, and report on, the state Seed Testing Laboratory;

☆ Advise on educational and promotional measures for proper enforcement and understanding of the Seeds Act;

☆ Plan for different varieties of crops to be grown in different regions of the state, and to review the assessment of seed requirements;

☆ Consider the release of new varieties for the state and recommend their notification to CSC

☆ Monitor the performance of newly released varieties in the state.

The SSSC is required to meet at least every quarter and plan strategies in cooperation with the State Seed Corporation and the Seed Certification Agency. An SSSC is responsible for the release of a variety only in its own state on the basis of data generated by the SAU. The state breeders, along with scientists of other disciplines, generate sufficient (usually three years) research trial data for establishing the VCU and other important features of the variety. On-farm trial data for one year are also collected by the extension agencies though this is not followed in all states. The SAU deliberates on the release proposal of a variety in a series of meetings before recommending it to the SSSC. Once approved by the SSSC for release in a state, the variety is required to be notified for seed production purposes by the Central Sub-Committee. A proposal for notification on the prescribed proforma (Annex 3) is then submitted by the SAU through SSSC. The notification proforma specifically requires that the variety must have been tested in the AICCIP trials at least for one year and preferably recommended for release in the state by the AICCIP Varietal Identification Committee.In a similar way, released varieties that are not performing satisfactorily in any region of a state, are not in much demand, or have been in cultivation for more than 15 years, are normally recommended for de-notification.

Inconsistencies between the Central and State Systems

Although they operate on different scales, both the central system and the system covering states have the same goal – of making the best material available to the farmers. But there are discrepancies between the two systems:

☆ Centrally released varieties are not automatically accepted by all states for which they are released.

In general, they have to pass through all the steps of the particular state release procedure before they are approved for cultivation in the state – multilocational trials within the state; research trials for three years; and adaptive trials (ATs) for one year where required. This procedure has two important implications.

   a. It is intended to check the spread of disease or pest susceptible varieties, but it can deprive other farmers of potentially beneficial genotypes.

b. State breeders may not favour the release in their state of a variety approved by CSC because it puts in question of their own output.

☆ While the State Seed Sub-Committee has power to release a variety, they have no authority to notify a variety for seed certification, even within the state: they are dependent on the central system for notification. The proforma for notification has a column that has to be filled in with details of a year's testing of the variety in All India Coordinated Crop Improvement Project trials, and the recommendation for its release within the state by the AICCIP workshop. This means that SSSCs have no independent purview: for a state variety to be notified, it has to have been through a much more rigorous procedure than centrally released varieties: firstly, rigorous multilocational testing within the state; secondly, testing across the country for the purpose of gaining identification for the state or zone within it.

This has often resulted in arguments between the central and the state organizations. It often takes a long time for state released varieties to be notified, and some are rejected for want of a recommendation from the AICCIP workshop. State releases may not reach farmers for want of notification, and the state may resort to the production of 'Truthfully Labelled' seed.

There has been a tendency in ICAR to impose restrictions on the state release system on the grounds that some states release an excessive number of varieties. Critics argue that this has resulted in too much centralization of the regulatory framework, and that this has the effect of restricting the basket of choices available to farmers.

Inconsistencies between State Systems

State-released varieties are constrained by a recommendation domain restricted to a single state because they are not accepted in other states with similar agro-ecological conditions. The way in which this can harm the interests of marginal farmers. In participatory trials on rice varieties in Rajasthan, Gujarat and Madhya Pradesh, it was found that all participating farmers preferred Kalinga III upland rice variety to the other varieties tested, because of its specific adaptation to low fertility and drought stress conditions. However, Kalinga III is only released in Orissa and not in any of the three states in question. For this variety to be promoted by state seed and extension agencies, it has to be released by all three State Seed Sub-Committees, a task so onerous for those promoting the variety that it is rendered impracticable. Therefore, a mechanism is required whereby a release proposal on the basis of data from farmers' field trials could be used for the zonal release of a variety, across an agro-ecological zone that covers more than one state.

Popularization

Following release, description of the new variety and recommendations for its production and protection are entered in a 'Package of Practices'. For central releases, the extension and seeds wings of Department of Agriculture and Cooperation, Ministry of Agriculture, seeds wing of ICAR, and AICCIP Directors/Coordinators are involved in the popularization of new varieties. Popularization programmes for newly released state varieties are planned by Directorates of Extension Education of SAUs, and state departments of agriculture in collaboration with agricultural universities. Education, exposure and seed is given to extension workers and farmers through brochures, hand-outs, demonstrations, training camps, trial visits, and *kissan melas* (farmer fairs) at farmers' fields.

An important role is played by the *Krishi Vigyan Kendras* (*KVK*) (farm science centres). KVKs demonstrate varieties on their campuses, comparing them with existing cultivars, and conduct front- line demonstrations on farmers' fields. At regional KVK *melas*, organised by agricultural universities, small quantities of the seed of new cultivars are sold to farmers, along with instructions about practices to be followed. A follow-up programme is organised and farmers' feedback is obtained for the performance of the variety.

However, evidence presented in the three subsequent chapters shows that information about modern varieties spreads slowly to farmers, and it takes several years after release before any significant extension activities are taken up by extension workers and seed producers. There is lack of communication about the release of new varieties since the information channel from the Central Sub- Committee through to seed producers and extension agencies to the farmers is very long.

Seed Production

The National Seeds Corporation and the State Farms Corporation of India (SFCI) undertake seed production programmes for a new variety using breeder seed supplied by the breeder. Seed multiplied at university farms is also supplied to state seed farms operated by the NSC and State Seeds Corporations. Seed multiplication takes place in the form of following categories:

☆ *Breeder seed:* Seed produced by the breeder of the variety from original or nucleus seed stocks.

Produced in small quantities.

☆ *Foundation seed:* Seed produced from breeder seed by selected growers under close supervision.

☆ *Certified seed:* Seed produced from foundation seed. Grown on a large-scale by seed organizations and farmers and sold for commercial crop production.

Seed growers, if they wish to produce certified seed, are required to get their seed certified by the State Seed Certification Agency (SSCA) if they wish to sell it as certified seed. In the case of hybrid seed, the parental seeds are supplied to registered producers, and strict control is kept on production of certified hybrid seed. Minimum seed certification standards were issued by the Central Seed Certification Board in July 1988 and cover, for example, inspections, minimum distances for isolation, objectionable seed, plants with seed borne diseases, and pollen shedders in male-sterile lines.

Some farmers multiply seed, maintaining sufficient purity, and sell it as Truthfully Labelled (TL) seed. For various reasons, including the considerable procedures involved in getting a variety released, some companies produce their own TL seed under their own trade name.

The non-availability of seed of new varieties appears to be a major constraint in the rapid doption of new varieties and replacement of old cultivars. New varieties do not enter into seed production for many seasons. Seed producers often do not place indents for new varieties immediately, since they are unable to estimate the seed demand from farmers who may be completely unaware of the new cultivars. Seed producers, to avoid risk, tend to estimate demand by asking farmers, rather than promoting unknown new cultivars.

APPENDICES

Appendix-J

**Proforma for Submission of Proposals for Identification of
Crop Varieties/Hybrids by the Workshops**

1. (a) Name of the Variety/Hybrid : _____

    (b) Species : _____

2. Parentage : _____

3. Breeding method used : _____

4. Developed by (Station and names of workers) : _____

5. Proposed by : _____

6. Zone for which to be identified : _____

7. Production condition for which : _____

8. In case of hybrid description of the parents : _____

9. List (at least two) important morphological : _____
    features of the proposed variety/hybrid which
    distinguish it from other important commercial
    varieties under field condition. Also enclose
    separately a complete description of the variety

10. The new variety/hybrid provides an alternative/ : _____
    replacement for

11. List main problems and special requirements (in
    order of importance of the concerned area of
    recommendation and how the proposed variety
    helps to resolve these : _____

12. Year when first entered in coordinated Varietal : _____
    Traits

13. Quantity of breeder seed available

    (a) Variety : _____

    (b) Parental lines in case of hybrids : _____

14. Summary/detailed data as per enclosures : _____

15. Problems and prospects in seed production of : _____
    parental lines and hybrids and their maintenance
    (wherever applicable)

Summary Yield Data of Coordinated Varietal Trails

Name of the proposed Variety/Hybrid: _____   Adaptability Zone: _____

Production _____

	Year of Testing	No. of Trials	Proposed Variety	Check Var.1	Check Var.2	Check Var.3	Qaul. Var.1	Qaul. Var.2	Qaul. Var.3	C.D.
Main yield (q/ha)										
a) Zonal	1st year									
	2nd year									
b) Across Zones (if applicable)	3rd year									
	Mean									
Percentage increase or decrease over the check and qualifying varieties	1st year									
	2nd year									
	3rd year									
Frequency in the top group (pooled for 3 years)										

Note:

1. Qualifying variety is one which has completed three years of testing in co-ordinated trails.

2. Centre-wise and year-wise data must be appended, otherwise proposal will not be considered.

Adaptability to Agronomic Variables

Name of the proposed Variety/Hybrid: _____   Adaptability Zone: _____

Production _____

Nature of Expt.	Item	Proposed Variety	Check1	Check 2	Check3	Qaul.Var.1	Qaul.Var.1	Qaul.Var.1
Sowing date experiments	Yield (Q/ha) under recommended sowing date							
	Percentage gain or loss when sown: i) Early ii) Normal iii) Late							
Fertilizer experiments	Yield (q/ha) under recommended dose							
	Percentage gain or loss under other doses i) F_0 ii) F_1 iii) F_2							
Irrigation experiments (wherever applicable)	Yield (Q/ha) with adequate irrigation							
	Percentage gain or loss with irrigation level i) Level 1 ii) Level 2 iii) Level 3							

Note: Specify each date of sowing, fertilizer level and number of irrigation at i, ii, iii, under column two.

Reaction to Major Diseases

Name of the proposed Variety/Hybrid: _____   Adaptability Zone: _____

Production _____

Disease Name	Item	Proposed Variety	Check Var.2	Check Var.2	Check Var.2	Qaul.Var.1	Qaul.Var.1	Qaul.Var.1
Disease (1)	NAT.	1st year 2nd year 3rd year						
	ART	1st year 2nd year 3rd year						
Disease (2)	NAT.	1st year 2nd year 3rd year						
	ART	1st year 2nd year 3rd year						
Disease (3)	NAT.	1st year 2nd year 3rd year						
	ART	1st year 2nd year 3rd year						
Disease (4)	NAT.	1st year 2nd year 3rd year						
	ART	1st year 2nd year 3rd year						

Disease Name	Item	Proposed Variety	Check Var.2	Check Var.2	Qaul.Var.1	Qaul.Var.1	Qaul.Var.1
Disease (5)	NAT.	1st year 2nd year 3rd year					
	ART	1st year 2nd year 3rd year					
Disease (6)	NAT.	1st year 2nd year 3rd year					
	ART	1st year 2nd year 3rd year					

NAT: Natural; ART: Artificial.

Reaction to Major Insect-Pests

Name of the proposed Variety/Hybrid: _____   Adaptability Zone: _____

Production _____

Insect Name	Item	Proposed Variety	Check Var.2	Check Var.2	Check Var.2	Qaul.Var.1	Qaul.Var.1	Qaul.Var.1	Qaul.Var.1
Pest (1)	NAT.	1st year 2nd year 3rd year							
	ART	1st year 2nd year 3rd year							
Pest (2)	NAT.	1st year 2nd year 3rd year							
	ART	1st year 2nd year 3rd year							
Pest (3)	NAT.	1st year 2nd year 3rd year							
	ART	1st year 2nd year 3rd year							
Pest (4)	NAT.	1st year 2nd year 3rd year							
	ART	1st year 2nd year 3rd year							

NAT: Natural; ART: Artificial.

Data on Quality Characteristics

Quality Characteristics	Item	Proposed Variety	Check Var.1	Check Var.2	Check Var.3	Qaul.Var.1	Qaul.Var.2	Qaul.Var.3
Parameter-1								
Parameter-2								
Parameter-3								
Parameter-4								
Parameter-5								

Note: Specify the parameter at 1 to 4 under first column

Appendix-JJ

Name of the Vegetables	Moisture (g)	Carbo hydrates (g)	Protein (g)	Fat (g)	Calorie (Energy)	Vit. A (IU)	Thiamine (mg)	Riboflavin (mg)	Ascorbic Acid (mg)	Calcium (mg)	Iron (mg)	Phosphorus (mg)
Potato	74.7	22.6	1.6	0.1	97	40	0.40	0.04	17.0	10.0	0.7	35.0
Tomato (green)	93.1	3.6	1.9	0.1	23	307	0.07	0.01	31.0	20.0	1.8	–
Tomato (ripe)	94.0	3.6	1.2	0.1	20	302	0.12	0.06	27.0	48.0	0.4	26.0
Chilli	85.7	3.0	2.9	0.6	29	292	0.19	0.39	111.0	30.0	1.2	80.0
Sweet pepper	92.4	4.3	1.3	0.3	24	683	0.06	0.03	175.0	10.0	1.2	–
Brinjal	92.7	4.0	1.4	0.3	24	118	0.04	0.11	12.0	18.0	0.9	47.0
Cabbage	92.4	5.3	1.4	0.2	29	80	0.06	0.05	100	46.0	0.8	38.0
Cauliflower	91.7	4.9	2.4	0.2	31	70	0.04	0.03	75.0	30.0	17.0	76.0
Knolkhol	90.1	6.7	2.1	0.1	36	20	0.05	0.10	50.0	20.0	0.4	60.0
Broccoli	89.9	5.5	3.3	0.2	37	3500	0.05	0.12	137.0	80.0	0.8	79.0
Brussels sprouts	84.9	8.9	4.4	0.5	58	400	0.05	0.13	105.0	40.0	1.6	80.0
Kale	86.6	7.2	3.9	0.6	50	7540	0.05	0.16	115.0	85.0	1.6	62.0
Wax gourd	96.5	1.9	0.4	0.1	10	0	0.06	0.01	1.0	30.0	0.8	20.0
Bitter gourd (large)	92.4	4.2	1.6	0.2	25	210	0.07	0.09	88.0	20.0	1.8	70.0
Bitter gourd (small)	83.2	10.6	2.1	1.0	60	210	0.07	0.06	96.0	23.0	2.0	38.0
Bottle gourd	96.1	2.5	0.2	0.1	12	0	0.03	0.01	6.0	20.0	0.7	10.0
Cucumber	96.1	2.5	0.4	0.1	13	0	0.03	0.01	7.0	10.0	1.5	25.0
Sponge gourd	93.2	2.9	0.5	0.1	18	120	0.02	0.06	0.0	36.0	1.1	19.0
Ridge gourd	95.2	3.4	0.5	0.1	17	56	0.01	0.01	5.0	18.0	0.5	26.0

Name of the Vegetables	Moisture (g)	Carbo hydrates (g)	Protein (g)	Fat (g)	Calorie (Energy)	Vit. A (IU)	Thiamine (mg)	Riboflavin (mg)	Ascorbic Acid (mg)	Calcium (mg)	Iron (mg)	Phosphorus (mg)
Snake gourd	94.6	3.3	0.5	0.3	18	160	0.04	0.06	0.0	26.0	0.3	20.0
Pointed gourd	92.0	2.2	2.0	0.3	20	255	0.05	0.06	29.0	30.0	1.7	40.0
Giant spine gourd	84.1	7.7	3.1	1.0	52	2592	0.05	0.18	0.0	33.0	4.6	42.0
Ivy gourd	93.5	3.1	1.2	0.1	18	249	0.07	0.08	15.0	40.0	1.4	30.0
Pumpkin (ripe)	86.0	4.6	1.4	0.1	25	2180	0.06	0.04	2.0	10.0	0.7	30.0
Summer squash	94.8	3.5	1.0	0.1	19	260	0.05	0.03	18.0	10.0	0.6	30.0
Winter squash (ripe)	86.0	6.6	1.1	0.2	32	3300	0.03	0.05	6.0	14.2	0.4	20.9
Muskmelon	95.2	3.5	0.3	0.2	17	3420	0.11	0.08	26.0	32.0	1.4	14.0
Watermelon	95.8	3.3	0.2	0.2	16	590	0.02	0.04	1.0	11.0	7.9	12.0
Round melon	93.5	3.4	1.4	0.2	21	23	0.04	0.08	18.0	25.0	0.9	24.0
Snap melon	95.7	3.0	0.3	0.1	14	265	–	–	10.0	–	–	–
Chowchow	92.5	5.7	0.7	0.1	27	8	0.01	0.04	4.0	140.0	0.6	30.0
Radish (white)	94.4	3.4	0.7	0.1	17	50	0.06	0.02	15.0	50.0	0.4	22.0
Radish (pink)	90.8	6.8	0.6	0.3	32	50	0.06	0.02	17.0	50.0	0.5	20.0
Carrot	82.2	10.6	0.9	0.2	48	12000	0.04	0.02	3.0	48.0	0.6	30.0
Beet	87.7	8.8	1.7	0.1	43	18	0.04	0.09	10.0	28.0	1.0	55.0
Parsnip	78.6	18.2	1.5	0.5	78	0	0.08	0.12	18.0	57.0	0.6	80.0
Turnip	91.6	6.2	0.5	0.2	28	4	0.04	0.04	43.0	30.0	0.4	40.0
Rutabaga	89.1	8.9	1.1	0.1	38	330	0.07	0.08	36.0	55.0	1.0	41.0
Onion	86.8	11.0	1.2	0.2	50	35	0.08	0.01	11.0	180.0	0.7	50.0
Garlic	62.8	29.0	6.3	0.1	142	10	0.16	0.23	13.0	30.0	1.3	310.0
Leek	78.9	5.0	1.8	0.1	28	30	0.23	–	11.0	–	2.3	70.0
Shallot	–	–	2.6	–	–	–	0.06	0.02	1.0	37.0	1.3	60.0

Name of the Vegetables	Moisture (g)	Carbo hydrates (g)	Protein (g)	Fat (g)	Calorie (Energy)	Vit. A (IU)	Thiamine (mg)	Riboflavin (mg)	Ascorbic Acid (mg)	Calcium (mg)	Iron (mg)	Phosphorus (mg)
Lettuce	93.4	2.5	2.1	0.3	21	540	0.09	0.13	10.0	50.0	2.4	28.0
Celery	88.0	1.6	6.3	0.6	38	0	0.01	0.11	7.0	50.0	0.6	38.0
Sorrel	95.2	1.4	1.6	0.3	14	–	0.03	0.06	12.0	63.0	8.7	–
Sweet corn	89.0	7.9	1.4	0.3	41	150	0.06	0.05	4.4	3.5	0.2	46.0
Okra	89.6	6.4	1.9	0.2	35	88	0.07	0.10	13.0	66.0	1.5	56.0
Pea	72.0	15.8	7.2	0.1	93	300	0.25	0.01	19.0	20.0	1.5	139.0
French bean	91.4	4.5	1.7	0.1	25	321	0.08	0.06	16.0	50.0	1.7	28.0
Cow pea	84.6	8.0	4.3	0.2	51	941	0.07	0.09	13.0	80.0	2.5	74.0
Hyacinth bean	86.1	6.7	3.8	0.7	48	312	0.1	0.06	9.0	210.0	1.7	68.0
Cluster bean	81.0	10.8	3.2	0.4	59	316	0.09	0.09	47.0	130.0	5.0	50.0
Winged bean	91.0	3.8	2.9	0.1	27	595	0.06	0.12	37.0	300.0	1.7	69.0
Broad bean	85.4	7.2	4.5	0.1	48	14	0.08	–	12.0	50.0	1.4	–
Lima bean	66.5	23.5	7.5	0.8	128	280	0.21	0.11	32.0	63.0	2.3	158.0
Sweet potato	70.0	27.0	1.5	0.2	115	2800	0.08	0.04	24.0	46.0	0.8	49.0
Taro corn	73.1	21.1	3.0	0.1	97	166	0.09	0.03	0.0	40.0	1.7	140.0
Giant taro corm	81.2	16.9	0.6	0.1	71	–	0.10	0.02	0.0	50.0	0.5	46.0
Tannia corm	75.4	21.0	2.2	0.1	93	–	0.12	0.04	8.0	34.0	1.2	62.0
Elephant foot yam	78.5	18.4	2.0	0.1	82	434	0.06	0.02	6.0	38.0	2.4	38.0
Yam root	73.0	22.0	2.7	0.3	109	–	0.09	0.03	5.0	40.0	1.7	–
Cassava	65.5	32.4	1.2	0.2	135	80	0.05	0.04	34.0	26.0	0.9	32.0
Chinese potato	77.6	19.7	1.3	0.1	85	–	–	–	–	–	–	–
Asparagus	91.7	2.9	1.7	0.2	20	762	0.12	0.13	25.0	15.8	0.7	46.9
Globe artichoke	83.7	11.9	2.9	0.4	63	390	0.18	0.05	10.0	47.0	1.9	94.0

Name of the Vegetables	Moisture (g)	Carbohydrates (g)	Protein (g)	Fat (g)	Calorie (Energy)	Vit. A (IU)	Thiamine (mg)	Riboflavin (mg)	Ascorbic Acid (mg)	Calcium (mg)	Iron (mg)	Phosphorus (mg)
Drumstick pod	86.9	3.7	2.5	0.1	25	176	0.05	0.07	120.0	30.0	3.3	110.0
Jackfruit (unripe)	84.0	9.4	2.6	0.3	50	0	0.05	0.04	14.0	30.0	1.7	–
Green banana	83.2	14.0	1.4	0.2	63	48	0.05	0.02	24.0	10.0	0.6	29.0
Green papaya	92.0	5.7	0.7	0.2	27	0	0.01	0.01	12.0	28.0	0.9	–
Green mango	87.5	10.1	0.7	0.1	44	144	0.04	0.01	3.0	10.0	–	–
Amaranth	85.7	6.3	4.0	0.5	45	9200	0.03	0.10	99.0	397.0	25.5	83.0
Palak	86.4	6.5	3.4	0.8	46	9770	0.26	0.56	70.0	380.0	16.2	30.0
Spinach	92.1	2.9	2.0	0.7	27	9300	0.03	0.07	28.0	73.0	10.9	21.0
Indian spinach	90.8	4.2	2.8	0.4	31	3250	0.03	0.16	87.0	200.0	1.0	35.0
Water spinach	92.4	–	1.9	–	–	4800	–	–	58.0	90.0	4.8	–
Fenugreek leaves	86.1	6.0	4.4	0.9	49	3744	0.05	–	54.0	360.0	17.2	51.0
Mustrad leaves	89.8	3.2	4.0	0.6	34	4195	0.03	–	33.0	155.0	16.3	26.0
Bengal gram leaves	73.4	14.1	7.0	1.4	97	1564	0.09	0.10	61.0	340.0	23.8	–
Bottle gourd leaves	87.9	6.1	2.3	0.7	40	–	–	–	–	80.0	–	59.0
Bottle gourd	81.9	7.9	4.6	0.8	57	5760	–	–	–	392.0	–	112.0
Pumpkin leaves	80.5	5.8	5.4	1.1	54	–	–	–	–	531.0	–	73.0
Pointed gourd leaves	86.3	6.3	3.3	0.6	44	11168	0.50	0.06	135.0	184.0	18.5	–
Coriander leaves	89.6	2.9	3.7	0.4	30	2784	0.01	0.14	35.0	150.0	4.2	–
Pigweed	90.5	2.7	2.4	0.6	26	3664	0.10	0.22	29.0	111.0	14.8	–
Portulacea	63.8	18.7	6.1	1.0	36	12096	0.08	0.01	4.0	830.0	7.0	57.0
Curry leaves	90.3	3.4	2.7	0.6	30	18	0.03	0.16	103.0	310.0	16.1	60.0
Radish top	83.9	8.3	5.1	0.5	58	–	–	–	–	340.0	8.3	110.0
Beet top	90.4	5.6	2.0	0.3	27	6700	0.08	0.18	35.0	118.0	3.1	45.0

Name of the Vegetables	Moisture (g)	Carbo hydrates (g)	Protein (g)	Fat (g)	Calorie (Energy)	Vit. A (IU)	Thiamine (mg)	Riboflavin (mg)	Ascorbic Acid (mg)	Calcium (mg)	Iron (mg)	Phosphorus (mg)
Agathi leaves	73.1	11.8	8.4	1.4	93	8640	0.21	0.09	58.0	404.0	5.0	–
Mint	84.9	5.8	4.8	0.6	48	2592	0.05	0.20	27.0	200.0	15.6	–
Ceylon spinach	93.6	4.0	1.6	–	22	5600	–	–	86.0	106.0	1.6	43.2
Swiss chard	91.8	4.4	1.6	0.2	21	3110	0.03	0.09	34.0	110.0	3.6	29.0
Cowpea leaves	89.0	4.1	3.4	0.7	20	9715	0.05	0.18	4.0	290.0	20.1	–
Drumstick leaves	75.9	12.5	6.7	1.7	92	10848	0.06	0.05	220.0	440.0	7.0	70.0
Chekkurmanis	73.6	11.6	6.8	3.2	102	9510	0.48	0.32	247.0	570.0	28.0	200.0
Onion scape	87.6	8.9	0.9	0.2	41	952	0.00	0.03	17.0	50.0	7.5	–
Green onion	87.6	10.6	1.0	0.2	45	50	0.03	0.04	24.0	135.0	0.9	24.0
Taro petioles	93.8	4.6	0.2	0.2	20	335	0.01	0.02	8.0	57.0	1.4	23.0
Chicory	94.2	2.9	1.6	0.3	21	10000	0.05	0.20	15.0	18.0	0.7	21.0
Chives	86.0	7.8	3.8	0.6	52	500	0.12	–	70.0	48.0	8.4	57.0
Water cress	93.6	3.3	1.7	0.3	18	4720	0.08	0.16	77.0	195.0	2.0	46.0
Endive	93.3	4.0	1.6	0.2	20	3000	0.07	0.12	11.0	79.0	1.7	56.0
Parsley	83.9	9.0	3.7	1.0	50	8230	0.11	0.28	193.0	193.0	4.3	84.0
Banana flower	89.9	5.1	1.7	0.7	33	43	0.05	0.02	16.0	32.0	1.6	–
Pumpkin flower	89.1	5.8	2.2	0.8	39	–	–	–	–	120.0	–	60.0
Agathi flower	87.4	–	1.8	0.6	–	–	0.13	–	41.0	–	–	–
Lotus flower	85.9	11.3	1.7	0.1	53	–	0.10	–	22.0	21.0	0.4	–
Mushrooms	91.1	4.0	2.4	0.3	16	0	0.10	0.44	5.0	9.0	1.0	115.0

Note: 0.6 µg β carotene = 1 IU vitamin A

– = data not available; 1 µg = 10^{-6} gram

Sources

1. Oser, L. Bernard (ed.) (1979), *Hawk's Physiological Chemistry*, Tata McGraw Hill.

2. Bose, T.K. and M.G. Som (eds) (1986), *Vegetable Crops in India*, Naya Prakash.

3. Thompson, H.C and W.C. Kelley, (1959), *Vegetable Crops*, Tata McGraw Hill.

4. Nath, Prem (1976), *Vegetables for the Tropical Region*, ICAR, New Delhi.

5. Ghosh, S.P., T. Ramanujam, J.S. Jos, S.N. Moorthy and R. G. Nair (1988) *Tuber Crops*, Oxford and IBH.

6. Konokv, P.F. and V. Kiran (1988), *Vegetable Growing in Home gardens of Tropical and Subtropical Areas*, Mir Pub., Moscow.

7. Peter, K.V. and V.S. Devdas (1989), Leafy Vegetables, *Indian Horticulture*, **33 and 34**: 8-11.

8. Shanmugavelu, K.G. (1993), *Production Technology of Vegetable Crops*, Oxford and IBH.

9. Indira, P and K.V. Peter, (1988), *Underexploited Tropical Vegetables*, Publication Unit, Directorate of Extension, Kerala Agricultural University.

Appendix-III

ICAR Agriculture Universities

(1) Acharya NG Ranga Agricultural University
Website: http: //www.angrau.net

Email: angrau_vc@yahoo.com, raghuvardhanreddy_s@rediffmail.com
Adminstrative Office, Rajendra Nagar, Hyderabad-500030, Andhra Pradesh 040-24015035, 24013095

Fax: 040-24015031

(2) Anand Agricultural University
Website: http: //www.aau.in

Email: vc@aau.in, vc_aau@yahoo.com Anand 388110, Gujarat 02692-261273

Fax: 02692-261520

(3) Assam Agricultural University
Website: http: //www.aau.ac.in

Email: vc@aau.ac.in, kmbujarbaruah@rediffmail.com Jorhat 785013, Assam 0376-2340013

Fax: 0376-2340001

(4) Bidhan Chandra Krishi Viswavidyalaya
Website: http: //www.bckv.edu.in

Email: ckvvc@gmail.com,sarojsanyal@yahoo.co.in Mohanpur, Nadia-741252, West Bengal 033-25879772, 03473-222666 Fax: 03473-222275

(5) Bihar Agricultural University
Website: http: //www.bausabour.ac.in

Email: vcbausabour@gmail.com Sabour,

Bhagalpur 813210, Bihar 0641-2452606Fax: 0641-2452604

(6) Birsa Agricultural University
Website: http: //www.baujharkhand.org
Email: vc_bau@rediffmail.com Kanke, Ranchi-834006, Jharkhand 0651-2450500

Fax: 0651-2450850

(7) Central Agricultural University
Website: http://www.cau.org.in

Email: snpuri04@yahoo.co.in, snpuri@rediffmail.com P.O. Box 23, Imphal-795004, Manipur 0385-2415933Fax: 0385-2410414

(8) Chandra Shekar Azad University of Agriculture and Technology
Website: http://www.csauk.ac.in

Email: vc@csauk.ac.in Kanpur-208002, Uttar Pradesh 0512-2534155

Fax: 0512-2533808

(9) Chaudhary Charan Singh Haryana Agricultural University
Website: http://www.hau.ernet.in

Email: vc@hau.ernet.in Hisar-125004, Haryana 01662-231640, 284301

Fax: 01662-234952

(10) CSK Himachal Pradesh Krishi Vishvavidyalaya
Website: http://www.hillagric.ac.in

Email: vc@hillagric.ac.in Palampur-176062, Himachal Pradesh 01894-230521

Fax: 01894-230465

(11) Chhattisgarh Kamdhenu Vishwavidyalaya
Website: http://cgkv.ac.in

Email: Anjora,Durg, **Chhattisgarh**

(12) Dr Balasaheb Sawant Konkan Krishi Vidyapeeth
Website: www.dbskkv.org

Email: vcbskkv@yahoo.co.in Dapoli Distt, Ratnagiri 415 712, Maharashtra 02358-282064

Fax: 02358-282074

(13) Dr Panjabrao Deshmukh Krishi Vidyapeeth
Website: http://www.pdkv.ac.in

Email: vc@pdkv.ac.in Krishinagar,Akola-444104, Maharashtra 0724-2258365

Fax: 0724-2258219

(14) Dr Yashwant Singh Parmar Univ of Horticulture and Forestry
Website: http://www.yspuniversity.ac.in

Email: vc@yspuniversity.ac.in, vcuhf@yahoo.com Solan, Nauni-173230, Himachal Pradesh 01792-252363

Fax: 01792-252242

(15) Dr YSR Horticultural University
Website: http: //www.drysrhu.edu.in

Email: vc@drysrhu.edu.in Adminstrative office, Venkataramannagudem, PB No. 7, West Godavari Dist., Tadepalligudem-534101, Andhra Pradesh 08818-284311

Fax: 08818-284223

(16) Govind Ballabh Pant University of Agriculture and Technology
Website: http: //www.gbpuat.ac.in

Email: vcgbpuat@gmail.com Pantnagar-263145, Distt Udham Singh, Nagar, Uttaranchal 05944-233333,233500

Fax: 05944-233500

(17) Guru Angad Dev Veterinary and Animal Science University
Website: http: //www.gadvasu.in

Email: vijay_taneja@hotmail.com, vcgadvasu@gmail.com Ludhiana - 141004, Punjab 0161-2553320,2553360

Fax: 0161-2553340

(18) Indira Gandhi Krishi Vishwavidyalaya
Website: www.igau.edu.in

Email: vcigkv@gmail.com Krishak Nagar, Raipur-492006, Chhattisgarh 0771-2443419

Fax: 0771-2442302, 2443121

(19) Jawaharlal Nehru Krishi Viswavidyalaya
Website: http: //www.jnkvv.nic.in

Email: gkalloo_jnkvv@yahoo.co.in Krishi Nagar, Adhartal

Jabalpur-482004, Madhya Pradesh 0761-2681706

Fax: 0761-2681389

(20) Junagadh Agricultural University
Website: http: //www.jau.in

Email: vc@jau.in Univ. Bhavan, Motibagh

Junagadh-362001, Gujarat 0285-2671784

Fax: 0285-2672004

(21) Karnataka Veterinary, Animal and Fisheries Sciences University
Website: http: //www.kvafsu.kar.nic.in

Email: vckvafsu@yahoo.co.in

dekvafsub@yahoo.com, sskvafsu@yahoo.co.in Nandinagar, PB No 6, BIDAR 585401, Karnataka 08482-245264

Fax: 08482- 245107

(22) Kerala Agricultural University
   Website: http://www.kau.edu

   Email: vc@kau.in, vicechancellorkau@gmail.com Vellanikara, Trichur 680656, Kerala 0487-2371928, 2370034, 2438001

   Fax: 0487-2370019

(23) Kerala University of Fisheries and Ocean Studies
   Website: http://www.kau.edu

   Email: kurup424@gmail.com Papangad, Kochi-682506 Kerala 0487-2370117, 2703781, 2700964

(24) Kerala Veterinary and Animal Sciences University
   Website: http://www.kcasu.ac.in

   Email: vc@kvasu.ac.in, vc.vetuny@gmail.com Liason Office, Directorate of Dairy

   Development, Pattom,Thiruvananthapuram, 695004, Kerala 0471-2550058

   Fax: 0471-2550480

(25) Lala Lajpat Rai University of Veterinary and Animal Sciences
   Website: http://www.llruvas.edu.in/

   Email: vc@llruvas.edu.in Hisar, Haryana 01662 289332

   01662- 272002

(26) Nanaji Deshmukh Veterinary Science University
   Website: http://www.mppcvv.org **Email:** vcnduvs@gmail.com South Civil Lines, Jabalpur-482001, Madhya Pradesh 0761-2678007

   Fax: 0761-2620783

(27) Maharana Pratap Univ. of Agriculture and Technology
   Website: http://www.mpuat.ac.in

   Email: vc@mpuat.ac.in Udaipur, Rajasthan 313001 0294-2471101

   Fax: 0294-2470682

(28) Maharashtra Animal Science and Fishery University
   Website: http://www.mafsu.in

   Email: vcmatsu@gmail.com, cadaba_prasad@yahoo.co.in Seminary Hills, Nagpur-440006, Maharashtra 0712-2511088

   Fax: 0712-2511282

(29) Mahatma Phule Krishi Vidyapeeth
   Website: http://mpkv.mah.nic.in

   Email: vcmpkv@rediffmail.com Rahuri-413722, Maharashtra 02426-243208

   Fax: 02426-243302

(30) Manyavar Shri Kanshiram Ji University of Agriculture and Technology
Website: http://www.mskjuat.edu.in/

Email: vc.mskjuat@gmail.com Banda - 210001,

Uttar Pradesh 05192-221605

Fax: 02426-243302

(31) Marathwada Agricultural University
Website: http://www.mkv2.mah.nic.in

Email: vcmau@rediffmail.com Parbhani-431402, Maharashtra 02452-223002

Fax: 02452-223582

(32) Narendra Deva University of Agriculture and Technology
Website: http://www.nduat.ernet.in

Email: vc_nduat2010@yaho.co.in Kumarganj, Faizabad -224229, Uttar Pradesh 05270-262097, 262161

Fax: 05270-262097

(33) Navsari Agricultural University
Website: http://www.nau.in

Email: vc_2004@yahoo.co.in Navsari-396450 Gujarat 02673-283869

Fax: 02673-284254

(34) Orissa Univ. of Agriculture and Technology
Website: http://www.ouat.ac.in

Email: ouat_dproy@yahoo.co.in, bsenapati1942@yahoo.com Bhubaneshwar-751003, Orissa 0674-2397700

Fax: 0674-2397780

(35) Punjab Agricultural University
Website: http://www.pau.edu

Email: vcpau@pau.edu Ludhiana-141004, Punjab 0161-2401794

Fax: 0161-2402483

(36) Rajasthan University of Veterinary and Animal Sciences
Website: http://rajuvas.org

Email: vcrajuvas@gmail.com Bijey Bhavan Place Complex

(Pt Deen Dayal circle) Bikaner 334006 Rajasthan

0151-2543419

Fax: 0151-2549348

(37) **Rajendra Agricultural University**
Website: http: //www.pusavarsity.org.in

Email: : vcrau@sify.com Pusa, Samastipur 848125, Bihar 06274-240226

Fax: 06274-240255

(38) Rajmata Vijayraje Sciendia Krishi Vishwa Vidyalaya
Website: http: //www.rvskvv.nic.in

Email: vcrvskvv@gmail.com Race Cource Road, Gwalior 474002 Madhya Pradesh 0751-2467673

Fax: 0751-2467673

(39) Sardar Vallabhbhai Patel University of Agriculture and Technology
Website: http: //www.svbpmeerut.ac.in

Email: vc_agunivmeerut@yahoo.com Modipuram, Meerut - 250110 Uttar Pradesh 0121-2888522, Fax: 0121-2888505

(40) Sardarkrushinagar-Dantiwada Agricultural University
Website: http: //www.sdau.edu.in

Email: vc@sdau.edu.in Sardar Krushinagar, Distt Banaskantha, Gujarat-385506 02748-278222

Fax: 02748-278261

(41) Sher-E-Kashmir Univ of Agricultural Sciences and Technology
Website: http: //www.skuast.org

Email: vc@skuast.org Railway Road, Jammu 18009, J and K 0191-2263714

Fax: 0191-2262073

(42)   Sher-E-Kashmir Univ of Agricultural Sciences and Technology of Kashmir
Website: http: //www.skuastkashmir.ac.in

Email: vc@skuastkashmir.ac.in, skuastkvc@gmail.com Shalimar Campus, Shrinagar-191121, Jammu and Kashmir 0194-2464028, 2462159

Fax: 0194-2462160, 2461543

(43) Sri Venkateswara Veterinary University
Website: http: //www.svvu.edu.in

Email: prabhakarvrao@yahoo.com Admn office, Regional Library Building, Tirupati-517502 0877-2248986

Fax: 0877-2249222

(44) Swami Keshwanand Rajasthan Agricultural University
Website: http: //www.raubikaner.org

Email: vcrau@raubikaner.org Bikaner-334006,Rajasthan 0151-2250443, 2250488

Fax: 0151-2250336

(45) Tamil Nadu Agricultural University
Website: http://www.tnau.ac.in

Email: vc@tnau.ac.in Coimbatore-641003, Tamil Nadu 0422-2431788, 2431672
Fax: 0422-2431672

(46) Tamil Nadu Fisheries University
Website: http://www.tnfu.org.in

Email: Nagapattinam – 611 003, Tamil Nadu 04365-253011

(47) Tamil Nadu Veterinary and Animal Science University
Website: http://www.tanuvas.ac.in

Email: vc@tanuvas.org.in, karanmgk@gmail.com Chennai-600051, Tamilnadu
044-25551574, 25551575
Fax: 0444-225551576

(48) University of Agricultural Sciences, Bangalore
Website: http://www.uasbangalore.edu.in

Email: vc@uasbangalore.edu.in,   knarayanagowda@yahoo.co.in
GKVK,Bengaluru-560065, Karnataka 080-23332442
Fax: 080-23330277

(49) University of Agricultural Sciences, Dharwad
Website: http://www.uasd.edu

Email: vc_uasd@rediffmail.com Dharwad-580005, Karnataka 0836-2447783,
Fax: 0836-2448349

(50) University of Agricultural Sciences, Shimoga
Website: http://www.uasbangalore.edu.in/asp/agriShimoga.asp

Email: Shimoga, Karnataka

(51) University of Horticultural Sciences
Website: http://uhsbagalkot.edu.in/

Email: dandinbnm@gmail.com Sector No 60 Navanagar Bagalkot 587102
Karnataka 08354 201310
Fax: 08354 235152

(52) University of Agricultural Sciences
Website: http://www.uasraichur.edu.in

Email: vcuasraichur10@rediffmail.com PB 329, Raichur – 584101 Karnataka
08532-221444
Fax: 08532- 220444

(53) UP Pandit Deen Dayal Upadhaya Pashu Chikitsa Vigyan Vishwa Vidhyalaya evam Go Anusandhan Sansthan
Website: http://www.upvetuniv.edu.in

Email: singhambika1945@gmail.com, duvasuvc@gmail.com Mathura-281001, Uttar Pradesh 0565-2470199

Fax: 0565-2404819

(54) Uttarakhand University of Horticulture and Forestry
Website: http://www.uuhf.ac.in/

Email: Pauri Garhwal, Uttarakhand 09412051976

(55) Uttar Banga Krishi Viswavidyalaya
Website: http://www.ubkv.ac.in

Email: vcubkv@gmail.com, vcubkv@rediffmail.com P.O. Pundibari, Dist. Coach Bihar-736165,West Bengal 03582-270141, 270013

Fax: 03582-270249

(56) West Bengal University of Animal and Fishery Sciences
Website: http://www.wbuafscl.ac.in

Email: wbuafs@wb.nic.in 68 KB Sarani, Kolkata-700037, West Bengal 033-25563450

Fax: 033-25571986

ICAR Institutions, Deemed Universities, National Research Centres, National Bureaux and Directorate/Project Directorates

Deemed Universities - 4

1. Indian Agricultural Research Institute, New Delhi
2. National Dairy Research Institute, Karnal
3. Indian Veterinary Research Institute, Izatnagar
4. Central Institute on Fisheries Education, Mumbai

Institutions - 47

1. Central Agricultural Research Institute, Port Blair
2. Central Arid Zone Research Institute, Jodhpur
3. Central Avian Research Institute, Izatnagar
4. Central Inland Fisheries Research Institute, Barrackpore
5. Central Institute Brackishwater Aquaculture, Chennai
6. Central Institute for Research on Buffaloes, Hissar
7. Central Institute for Research on Goats, Makhdoom
8. Central Institute of Agricultural Engineering, Bhopal
9. Central Institute of Arid Horticulture, Bikaner

10. Central Institute of Cotton Research, Nagpur
11. Central Institute of Fisheries Technology, Cochin
12. Central Institute of Freshwater Aquaculture, Bhubneshwar
13. Central Institute of Research on Cotton Technology, Mumbai
14. Central Institute of Sub Tropical Horticulture, Lucknow
15. Central Institute of Temperate Horticulture, Srinagar
16. Central Institute on Post harvest Engineering and Technology, Ludhiana
17. Central Marine Fisheries Research Institute, Kochi
18. Central Plantation Crops Research Institute, Kasargod
19. Central Potato Research Institute, Shimla
20. Central Research Institute for Jute and Allied Fibres, Barrackpore
21. Central Research Institute of Dryland Agriculture, Hyderabad
22. Central Rice Research Institute, Cuttack
23. Central Sheep and Wool Research Institute, Avikanagar, Rajasthan
24. Central Soil and Water Conservation Research and Training Institute, Dehradun
25. Central Soil Salinity Research Institute, Karnal
26. Central Tobacco Research Institute, Rajahmundry
27. Central Tuber Crops Research Institute, Trivandrum
28. ICAR Research Complex for Eastern Region, Patna
29. ICAR Research Complex for NEH Region, Barapani
30. ICAR Research Complex for Goa, Ela, Old Goa, Goa
31. Indian Agricultural Statistics Research Institute, New Delhi
32. Indian Grassland and Fodder Research Institute, Jhansi
33. Indian Institute of Agricultural Biotechnology, Ranchi
34. Indian Institute of Horticultural Research, Bengaluru
35. Indian Institute of Natural Resins and Gums, Ranchi
36. Indian Institute of Pulses Research, Kanpur
37. Indian Institute of Soil Sciences, Bhopal
38. Indian Institute of Spices Research, Calicut
39. Indian Institute of Sugarcane Research, Lucknow
40. Indian Institute of Vegetable Research, Varanasi
41. National Academy of Agricultural Research and Management, Hyderabad
42. National Biotic Stress Management Institute, Raipur
43. National Institue of Abiotic Stress Management, Malegaon, Maharashtra
44. National Institute of Animal Nutrition and Physiology, Bengaluru
45. National Institute of Research on Jute and Allied Fibre Technology, Kolkata

46. Sugarcane Breeding Institute, Coimbatore
47. Vivekananda Parvatiya Krishi Anusandhan Sansthan, Almora

National Research Centres - 17

1. National Centre for Agril. Economics and Policy Research, New Delhi
2. National Centre for Integrated Pest Management, New Delhi
3. National Research Centre for Agroforestry, Jhansi
4. National Research Centre for Banana, Trichi
5. National Research Centre for Citrus, Nagpur
6. National Research Centre for Grapes, Pune
7. National Research Centre for Litchi, Muzaffarpur
8. National Research Centre for Pomegranate, Solapur
9. National Research Centre on Camel, Bikaner
10. National Research Centre on Equines, Hisar
11. National Research Centre on Meat, Hyderabad
12. National Research Centre on Mithun, Medziphema, Nagaland
13. National Research Centre on Orchids, Pakyong, Sikkim
14. National Research Centre on Pig, Guwahati
15. National Research Centre on Plant Biotechnology, New Delhi
16. National Research Centre on Seed Spices, Ajmer
17. National Research Centre on Yak, West Kemang

National Bureaux - 6

1. National Bureau of Plant Genetics Resources, New Delhi
2. National Bureau of Agriculturally Important Micro-organisms, Mau, Uttar Pradesh
3. National Bureau of Agriculturally Important Insects, Bengaluru
4. National Bureau of Soil Survey and Land Use Planning, Nagpur
5. National Bureau of Animal Genetic Resources, Karnal
6. National Bureau of Fish Genetic Resources, Lucknow

Directorates/Project Directorates - 25

1. Directorate of Maize Research, New Delhi.
2. Directorate of Rice Research, Hyderabad
3. Directorate of Wheat Research, Karnal
4. Directorate of Oilseed Research, Hyderabad
5. Directorate of Seed Research, Mau
6. Directorate of Sorghum Research, Hyderabad
7. Directorate of Groundnut Research, Junagarh

8. Directorate of Soybean Research, Indore
9. Directorate of Rapeseed and Mustard Research, Bharatpur
10. Directorate of Mushroom Research, Solan
11. Directorate on Onion and Garlic Research, Pune
12. Directorate of Cashew Research, Puttur
13. Directorate of Oil Palm Research, Pedavegi, West Godawari
14. Directorate of Medicinal and Aromatic Plants Research, Anand
15. Directorate of Floricultural Research, Pusa, New Delhi
16. Project Directorate for Farming Systems Research, Modipuram
17. Directorate of Water Management Research, Bhubaneshwar
18. Directorate of Weed Science Research, Jabalpur
19. Project Directorate on Cattle, Meerut
20. Project Directorate on Foot and Mouth Disease, Mukteshwar
21. Directorate of Poultry Research, Hyderabad
22. Project Directorate on Animal Disease Monitoring and Surveillance, Hebbal, Bengaluru
23. Directorate of Knowledge Management in Agriculture (DKMA), New Delhi
24. Directorate of Cold Water Fisheries Research, Bhimtal, Nainital
25. Directorate of Research on Women in Agriculture, Bhubaneshwar

Krishi Vigyan Kendras (KVKs)

Krishi Vigyan Kendras	*No. of KVKs*
Zone I – 69 KVKs	
Delhi	1
Haryana	18
Himachal Pradesh	12
Jammu and Kashmir	18
Punjab	20
Zone II – 81 KVKs	
A and N Islands	3
Bihar	38
Jharkhand	22
West Bengal	18

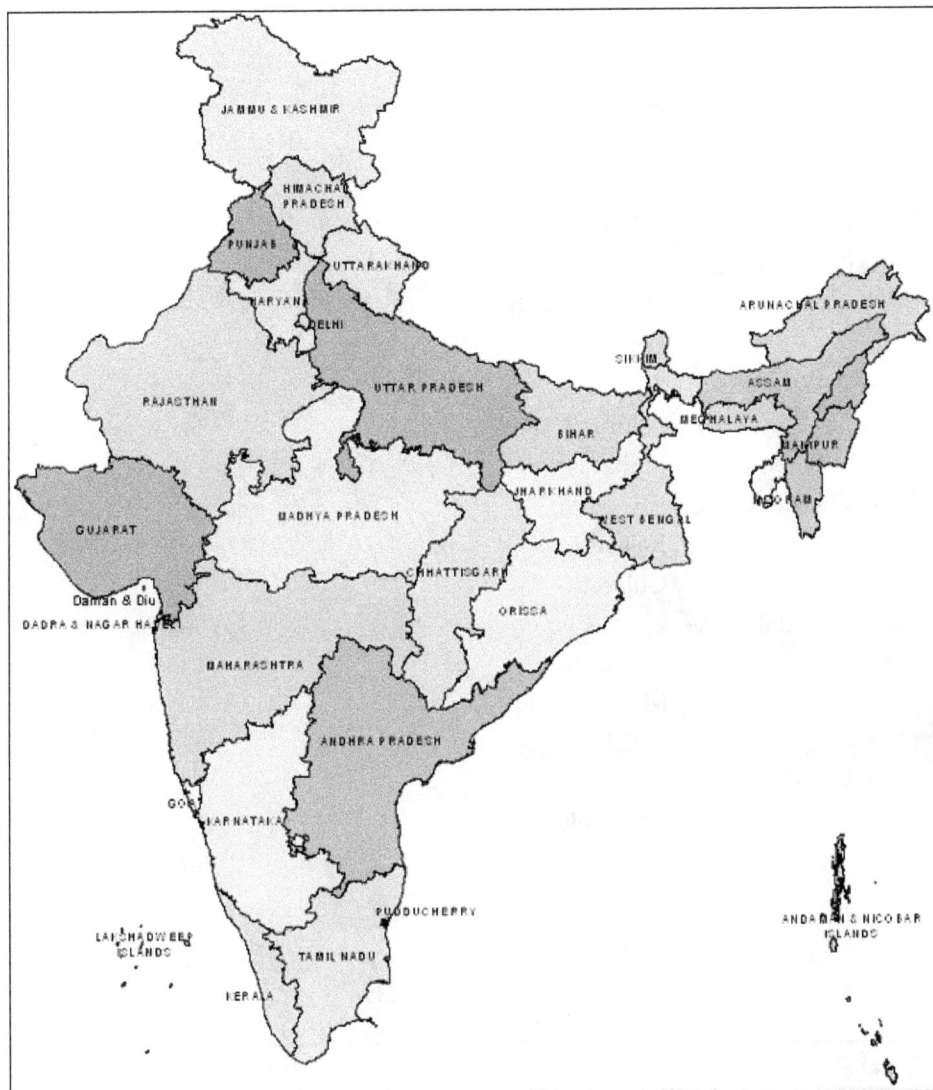

Krishi Vigyan Kendras (KVKs)

Krishi Vigyan Kendras		No. of KVKs
Zone III – 74 KVKs		
	Assam	22
	Arunachal Pradesh	13
	Manipur	9
	Meghalaya	5
	Mizoram	8
	Nagaland	9
	Sikkim	4
	Tripura	4
Zone IV – 81 KVKs		
	Uttar Pradesh	68
	Uttarakhand	13
Zone V – 78 KVKs		
	Andhra Pradesh	34
	Maharashtra	44
Zone VI - 70 KVKs		
	Rajasthan	42
	Gujarat	28
Zone VII – 100 KVKs		
	Chattisgarh	20
	Madhya Pradesh	47
	Odisha	33
Zone VIII – 81 KVKs		
	Karnataka	31
	Tamil Nadu	30
	Kerala	14
	Goa	2
	Pondicherry	3
	Lakshadweep	1
Total		**634**

ICAR AWARDS

ICAR Norman Borlaug Award

In order to recognize a scientist, who has provided a breakthrough for agriculture through a new insight that has created high potential value for the future the Norman Borlaug Award has been constituted. The nominations for the awards are invited in the prescribed pro-forma for a scientist(s) of any discipline of agricultural and allied sciences and not necessarily confined to NARS, whose research work has displayed a rare quality of original thinking and path-breaking output. The award would be of **Rs.10 lakh** in cash. Apart from this, the selected scientist would be given a research contingency grant of **Rs.30 lakh** per year for carrying out research in an area identified mutually by the scientist and the ICAR, which will have specified objectives and goals. Grant would be admissible for a maximum of five years, subject to annual review and favorable recommendation by a committee of agricultural and non-agricultural scientists set up for the project. The maximum age of the scientist for the award has been fixed as 50 years at the time of conferring of the award *i.e.* on **16th July 2014**.

ICAR Challenge Award

To find a solution for any immediate or long-standing problem, or limitation in agriculture, which is coming in the way of agricultural development enhancing productivity in any major agricultural, horticultural or animal/fish product, ICAR has instituted Challenge Award. The award shall consist of **Rs.10 lakh** in cash and a share in the income to ICAR from the commercialization as per rules of ICAR. The list of "Challenges" can be obtained either from the Award Cell of ICAR or can be down loaded from ICAR website. Any scientist or group of scientists may, at any time, file a claim of having solved any one of these "challenges".

Challenges

Eradication of bacterial blight in Pomegranate.

Development a sugarcane variety fully resistant to red rot.

Developing a variety of pulse crop, fully resistant to pod borer

RNA interference (RNAi) based cure for foot and mouth diseases (FMD).

Nano particle based prophylactic for avian influenza (AI) in poultry.

Development of yellow rust resistant (including ug 99) wheat.

Dev of HYV rice (>5 ton/ha) in conditions of low light intensity in the Eastern region.

Sardar Patel Outstanding ICAR Institution Award

In order to recognize outstanding performance by the ICAR institutes, DUs of ICAR, CAU and State Agricultural Universities, three Awards of **Rs.10.00 lakh** each, will be given to two ICAR Institutes/NRC/Project Directorates National Bureaus (One to Large and other to small) and one to State Agricultural University/DUs/CAU.

Chaudhary Devi Lal Outstanding All India Coordinated Research Project Award

In order to recognize outstanding performance of the AICRP and its cooperating centers and to provide incentive for outstanding performance in terms of linkages and research output and its impact, one annual award of **Rs.3 lakh (Rs. 2 lakh for the main coordinating unit and Rs. 1 lakh for the best centre)** in cash is to be given away. All the All India Coordinated Research Projects, which have been in operation for at least 10 years can apply for the award.

Jawaharlal Nehru Award for P.G. Outstanding Doctoral Thesis Research in Agricultural and Allied Sciences

In order to promote high quality doctoral thesis research in priority/frontier areas of agriculture and allied sciences, ICAR has instituted **18** awards of **Rs.50, 000/ -each** to be given away annually for the outstanding original research work in agriculture and allied sciences. This award is meant exclusively for doctoral thesis related to agricultural sciences from any Indian University. The Certificate of the Ph.D. degree/provisional degree must have been obtained during the year preceding the year of the award *i.e.***2012**. A thesis will be considered only once. Applicant must have published/accepted at least one good research paper in a journal having NAAS rating of ≥ 6 and above from the work done for the thesis.

Panjabrao Deshmukh Outstanding Woman Scientist Award

All women scientists engaged in research in agricultural and allied subjects/ extension in a recognized institutions are eligible The award consists of **Rs.100,000** in cash and citation along with provision of equal amount of **Rs. 1 lakh** for motivating Woman Scientists and female student across the country including travel within a year of receiving the award. The awards are exclusively meant for individual women scientists.

Vasantrao Naik Award for Outstanding Research Application in Dry Land Farming Systems

In order to promote outstanding research and application in priority aspects of dry land farming systems and water conservation, an Annual Award of **Rs. 1,00,000** is to be given away to a scientist or extension worker. All agricultural and allied scientists engaged in research/technology application work in dry land farming in India are eligible for the award.

Jagjivan Ram Abhinav Kisan Puruskar/Jagjivan Ram Innovative Farmer Award (National/Zonal)

In order to recognize the outstanding contributions of innovative farmers for initiatives in development adoption, modification and dissemination of improved technology and practices for increased income with sustainability, following national and zonal awards are announced:

National: One annual national award of **Rs. 1,00,000** each in any of the areas of agriculture and allied sciences + equal amount of travel grant across the country to promote his achievement are to be given to farmers at national level. Agricultural Production Commissioners/Secretaries/Directors of Agril./Hort./A.H./Fisheries/ sericulture, Vice-Chancellors of agricultural universities/Directors of ICAR Institute will identify and nominate the farmers in their area of jurisdiction and forward the authentic information to the Council.

Zonal: Eight annual awards of **Rs. 0.50 lakh** + equal amount of travel grant to promote his achievement and motivate farmers in his perspective zone. All the KVKs in the country are divided zone wise. There are eight zones and each zone is headed by Zonal Project Director. The geographical area of each zone is given in the guidelines of award. Agricultural Production Commissioners/Secretaries/Directors of Agril./ Hort/A.H./Fisheries/Sericulture, Vice-Chancellors of agricultural universities/ Directors of ICAR Institutes and Zonal Directors of KVKs/NGOs, will identify and nominate the farmers in their area of jurisdiction and forward the authentic information to the Council.

N.G. Ranga Farmer Award for Diversified Agriculture

In order to recognize outstanding contribution of innovative farmers for diversified agriculture, one annual award of **Rs. 1,00,000** in any of the areas of Diversified Agriculture is given by ICAR. Agricultural Production Commissioners/ Secretaries/Directors of Agril./Hort./A.H./Fisheries/Sericulture Vice-Chancellors of AUs/Directors of ICAR Instt./Zonal Directors of KVKs/NGOs, other organizations connected with plant and animal sciences will identify and nominate the farmers in their area of jurisdiction and forward the authentic information to the Council. All farmers selected for Zonal award and meeting the criteria of this award are also eligible.

Chaudhary Charan Singh Award for Excellence in Journalism in Agricultural Research and Development

In order to recognize the outstanding contribution in Journalism in the field of Agricultural Research and Development in the country, two annual award of **Rs. 1,00,000** are to be given to journalists. There will be two awards- one each in print and electronic media. The contribution made by the journalist would be judged through articles/success stories in Hindi/English News papers/Magazines/Journals/ electronic media in India during the preceding three years.

Fakhruddin Ali Ahmed Award for Outstanding Research in Tribal Farming Systems

The award is primarily meant for any person or team (with two or three associates, if any) engaged in applied research and its applications in tribal areas of the country aimed at improving the biological resources and livelihoods or in original work directly applicable to tribal farming system. Two awards of the value of **Rs.1,00,000 (One Lakh Only)** in cash and citation + Provision of equal amount for study on related subject in geographical area for a year.

Bharat Ratna Dr. C. Subramaniam Award for Outstanding Teachers

In order to provide recognition to outstanding teachers and to promote quality teaching, four awards are to be given annually. Each award consists of **Rs.1,00,000 in cash + travel grant of Rs. 1.0 lakh** to promote innovation in teaching. The Applications through proper channel of eligible teachers are invited from SAUs, CAUs and DUs.

Best Krishi Vigyan Kendra Awards(National and Zonal)

These awards recognize for outstanding performance by Krishi Vigyan Kendra at national level and zonal level and provide incentives for outstanding KVK performance, promote a sense of institutional pride in KVK for developing models of Extension Education and Technology application. For National level there is one award comprising of **Rs. 3.0 lakh + 3 lakh** for infrastructural development +**1.0 lakh** for sharing among staff + **5.0 lakh** for overseas training of Programme Coordinator. At zonal level there are total eight awards: one for each zone of KVKs. Each award consists of Rs.**1.0 lakh + 2.0 lakh** for infrastructural development + **1.0 lakh** for sharing among staff + **1.0 lakh** for training in Indian Institute for Programme Coordinators.

Dr. Rajendra Prasad Puruskar for technical books in Hindi in Agricultural and Allied Sciences

These awards recognize to authors of original Hindi Technical books in agriculture and allied sciences and incentivize Indian writers to write original standard works in agricultural and allied sciences in Hindi. The award is meant for individuals as well as teams of authors. An individual award consists of **Rs.1,00,000** in cash. Four awards, one each in the crop/Horticultural Sciences; NRM/Agricultural Engineering, Animal/Fisheries sciences and one Social Sciences. All original Hindi technical books in the designated subject areas of agriculture and allied sciences written by Indian authors, including editors of multi-author books in which the editor also has himself contributed substantially in, are eligible. The author/editor must have had a substantial and active involvement in the relevant field of agriculture and allied sciences. The publication must be free from any infringement of copy rights. The publication must have been written and published during preceding year of the award.

Lal Bahadur Shastri Outstanding Young Scientist Award

In order to recognize the talented young scientists who have shown extraordinary originality and dedication in their research programmes, four individual awards are to be given? annually. An individual award of Rs 100,000 in cash and a challenge project for three years with budgetary provision **of Rs. 10.0 lakh per year+ Rs. 5.0 lakh** for foreign training (3 months) if it is necessary by the ICAR. The challenge project and foreign training will be administered/monitored by Division of Agricultural Education at ICAR, Headquarters. All young scientists who possess a doctoral degree and are below forty years of age (on 31st Dec.2012) and hold a regular teaching, research, extension education job in the ICAR-SAU system of institutions and engaged in research in agricultural and allied sciences for at least five years continuously are eligible for consideration.

Rafi Ahmed Kidwai Award for Outstanding Research in Agricultural Sciences

In order to recognize outstanding research in agricultural and allied sciences and provide incentives for excellence in agricultural research, this award is to be given to agricultural scientists for outstanding contribution in specified areas. A total of four awards are provided under the award. Each award consists of **Rs. 5,00,000** in cash. All Indian scientists engaged in agricultural research and overseas Indian scientists working in the areas relevant to Indian agriculture are eligible for these awards.

Swami Sahajanand Saraswati Outstanding Extension Scientist Award

The award is exclusively meant for individual extension scientist/teacher for excellence in agricultural extension methodology and education work. Two individual awards have been provided. An individual award would consist of **Rs. 100,000/-** in cash. Two awards have been assigned across the disciplines in agriculture and allied sciences.

ICAR Award for Outstanding Interdisciplinary Team Research in Agricultural and Allied Sciences

To recognize encourage and promote the understanding that practical and useful research would normally have to be interdisciplinary approach. It is exclusively meant for interdisciplinary team of scientists jointly planning and implementing integrated programme/project. There would be maximum of 4 awards, and each award would be of Rs. 5,00,000/-. All agricultural scientists engaged in interdisciplinary team research in India in the specified subject areas are eligible.

Appendix IV

Conversion Factors Useful in Field Work

Area

Hectare	= 2.4711
	= 10,000 m^2
Acre	= 0.405 hectare
	= 4047 m^2

Volume

Litre	= 0.22 Imperial gallon
	= 0.2642 US gallon
	= 1.05 US quarts
	= 0.0353 cu ft
Litre/ha	= 0.107 US gallon/acre
Gallon (Imperial)	= 4.456 litres
	= 9.815 pounds water
Gallon (US)	= 3.785 litres
	= 8.3370 pounds of water
	= 4 quarts = 4 pints = 16 cups
Ounce (Imperial)	= 28.41 milliliters
Ounces (US)	= 29.573
	= 28.35 grams

Weight

kg (Kilogram)	= 1000 grams
	= 2.205 pounds = 35.27 ounces
Pound	= 453.59 grams
	= 0.4536 kg = 16 ounces

Length

Centimeter	= 0.3937 inch
Meter	= 3.28 feet = 39.4 inches
Kilometer	= 0.621 mile
Inch	= 2.54 centimeters
Foot	= 30.48 centimeters
Meter	= 39.37 inches
Yard	= 0.914 meter
Mile	= 1.61 kilometers
Square inch	$= 6.452 \text{ cm}^2$
Square foot	$= 0.093 \text{ m}^2$
Square meter	= 10.76 sq ft
Square kilometer	= 100 hectares
Square mile	= 620 acres = 260 hectares
	= 2.6 sq km
Degree Centigarde	= (F -32) x 5/9
Degree Farenheit	= (C x 9/5) + 32

Pressure

Pound per sq in (psi)	$= 0.700 \text{ kg/cm}^2 = 700 \text{ g/cm}^2$

Others

Foot candle	= 10.764 lux
Acre foot	= 1219.85 cu meters = 0.122 hectare meter
	= 1.23 million L water
Cu ft	= 28.4 L water = 0.0284 cu meter
Cu m water	= 0.0001 hectare meter water = 1010.7 kg water
Cusec	= 0.028 cubic m/sec (cum sec)
Pound/gallon	= 0.12 kg/L
Pound/acre (U.S.)	= 1.12 kg/ha

Appendix-V

Name of the Fertilizer	N per cent (N_2)	P per cent (P_2O_5)	K per cent (K_2O)
Urea (biuret)	46	–	–
Urea (Coated)	45	–	–
Ammonium sulphate	16	–	–
Ammonium chloride	25	–	–
Calcium ammonium nitrate (CAN)	26	–	–
Diammonium phosphate	18	46	–
Single super phosphate	–	46	–
Triple super phosphate	–	46	–
Bone meal raw	3	20	–
Bone meal steamed	–	20	–
Rock phosphate	–	26	–
Potassium nitrate	14	–	46
Potassium chloride (Muriate of potash)	–	–	60
Potassium sulphate	–	–	48
Ammonium phosphate	28	28	–
Diammonium phosphate (DAP)	18	46	–
Fertilizer mixtures	15	15	15
Fertilizer mixtures	17	17	17
Fertilizer mixtures	18	18	09
Fertilizer mixtures	19	19	19
Fertilizer mixtures	14	35	14
Fertilizer mixtures	10	26	26
Fertilizer mixtures	12	32	16

Name of the Amendment	N	P_2O_5	K_2O
Manure			
Compost (rural)	0.5-1.0	0.4-0.8	0.8-1.2
Compost (urban)	0.7-2.0	0.9-3.0	1.0-2.0
Farmyard manure	0.4-1.5	0.3-0.9	0.8-1.0
Goat and sheep manure	0.6-0.8	0.3-0.5	0.6-0.8
Poultry manure	1.0-1.8	1.4-1.8	0.8-0.9
Sewage sludge, dry	4.0-7.0	2.1-4.2	0.5-0.7
Wood ash	Traces	2.5	5.0-25.0
Green manure			
Dhaincha	0.62	–	–
Cowpea	0.71	0.15	0.58
Black gram	0.85	0.18	0.53
Clusterbean	0.34	–	–
Green gram	0.72	0.18	0.53
Gliricidia	2.90	2.40	2.40
Sunnhemp	0.75	0.12	0.51
Straw and Stalks			
Bajra	0.65	0.75	2.50
Banana, dry	0.61	0.12	1.00
Cotton	0.44	0.10	0.66
Jowar	0.40	0.23	2.17
Maize	0.42	1.57	1.65
Paddy	0.36	0.08	0.71
Tobacco	1.12	0.84	0.80
Tur, Arhar	1.10	0.58	1.28
Sugarcane trash	0.35	0.10	0.60
Wheat straw	0.53	0.10	1.10
Dry leaves of trees			
Calotropis gigantea	0.35	0.12	0.36
Careya arborea	1.67	0.40	2.20
Cassia auriculata	0.98	0.12	0.67
Dillenia pentagyna	1.34	0.50	2.00
Madhuca indica	1.66	0.50	2.00
Pongamia pinnata	3.69	2.41	2.42
Pterocarpus marsupium	1.97	0.40	2.90
Terminalia chebula	1.46	0.35	1.35
Terminalia paniculata	1.70	0.40	1.60
Terminalia tomentosa	1.39	0.40	1.80
Xylia dolabriformis	1.37	0.30	1.61

Name of the Amendment	N	P_2O_5	K_2O
Oil cakes			
Caster cake	4.0-4.4	1.9	1.4
Coconut cake	3.4	1.5	2.0
Cotton seed cake			
Decorticated	6.9	3.1	1.6
Undecorticated	3.6	2.5	1.6
Groundnut cake	6.5-7.5	1.3	1.5
Jambo cake	5.0	1.7	1.9
Karanj cake	4.0	0.9	1.3
Linseed cake	4.7	11.7	1.3
Mahua cake	2.5	0.8	1.9
Neem cake	5.2-5.6	1.1	1.5
Niger cake	4.7	1.8	1.3
Palmnut cake	2.6	1.1	0.5
Rape cake	4.8	2.0	1.3
Safflower cake			
decorticated	7.9	2.2	1.9
undecorticated	4.9	1.4	1.2
Sesamum cake	4.7-6.2	2.1	1.3

Source: Handbook of Agriculture.

Appendix-VI

Residues of some of the Herbicides Permitted Limit

Herbicide	Upper Residue Limit (ppm)	Herbicide	Upper Residue Limit (ppm)
Alachlor	0.2	Flumeturon	0.10
Ametryn	0.25	Linuron	0.10
Atrazin	0.25	Monuron	1.00
Barban	0.10	Nitralin	0.10
Bromacil	0.10	Nitrofen	0.75
Butylate	0.10	Noruron	0.20
Chloramben	0.10	Paraquat	0.05
Chloroxuron	0.15	Prometryne	0.25
Dalapon	10.00	Propachlor	0.10
Dicamba	0.50	Propazine	0.25
Dichlobenil	0.15	Simazine	0.25
Diuron	1.0	Terbacil	0.10
2,4-D	6.0	Trifluralin	1.00
EPTC	0.1	Vernolate	0.10

Source: NAC News, February 1970, National Agricultural Chemicals Association, USA.

Fertilizers and its Relation of Plant Nutrients

1 kg Nitrogen	= 5 kg Ammonium Sulphate
	or 4 kg Calcium Ammonium Nitrate (CAN)
	or 4 kg Ammonium Chloride
	or 2.22 kg Urea
	or 4 Kg Ammonium Sulphate Nitrate
1 kg Phosphate	= 6.25 kg Super Phosphate (Single)
1 kg Potash	= 1.660 kg Muriate of Potash (60 per cent) or
	Or 2 kg Muriate of Potash (50 per cent)

1 kg Nitrogen + 1 kg Phosphate

= 5 kg Suphala 20:20:0

1 kg itrogen + 1 kg Phosphate

= 3.5 kg Urea Ammonium Phosphate

1 kg Nitrogen + 1 kg Phosphate + 1 kg Potash

= 6.660 kg Suphala 15:15:15

Glossary

Abaxial: Leaf surface facing away from the stem (lower surface).

Abiotic pollination: Transfer of pollen by a biotic means, such as wind (*Anemophily*), water (*Hydrpphily*), on the water surface (*Ephydophily*), under the water surface (*Hyphydropily*) and gravity (*Geitonogam*).

Abnormal seedling: Seedling which does not show the capacity for continued development into normal plant under favourable conditions.

Abortive: Defective, barren or imperfectly developed (Embryo, pollen etc.).

Abscission: Shedding of plant parts as a result of the formation of an abscission layer of loosely adhering cells at its base which breaks apart readily; regulated by abscissic acid.

Absolute growth rate: A growth parameter depicting the rate of increase in size of a growing plant (or part of it) on a given time under specific condition, expressed as dry weight/time.

Absorption spectrum: Relative absorbance of various wave-lengths of light by a compound such as chlorophyll, measured by spectrophotometer.

Acclimatization: Gradual process of establishment of a plant in a new locality in which environmental conditions differ markedly from those of its native habitat.

Acicular: Long, narrow and cylindrical, *i.e.*, needle- shaped as the leaves of onion, pine etc.

Acid foods: Foods having pH 4.5 to 3.7 which are usually spoiled by non-spore forming aciduric, butyric anaerobes, etc. *e.g.* products of tomato, pear, etc.

Acre: A unit of land measures equal to 43, 560 sq. ft.; 4,840 sq. yds; 4,047 sq. mts or 0.4 hectares.

Acre inch: A measure of quantity of water flow covering an acre to a depth of one inch, assuming no seepage, evaporation and run off loses.

Acridity: Sense of irritation in a testing different aroids due to the presence of needle like crystals of calcium oxalate in the different plant parts of the aroids (Taro, Elephant foot yam etc.).

Acropetal: Development of organs in succession towards the apex, the oldest at base, youngest at tip, *e.g.* leaves on a shoot.

Active absorption: absorption of water and other substances against the concentration gradient, an energy requiring processes.

Active ingredient: The active component of a formulated product such as fungicide, insecticide, herbicide etc.

Adsorption: The tendency of molecules or ions to adhere to the surface of certain solids in liquids.

Adventitious buds: Buds or shoots arising from any plant part other then terminal, lateral, latent buds on stems.

Adventitious root: Roots arising from any plant aprt other than the seedling root and its branches.

Aeration (Soil): The process by which air and other gases in the soil are renewed which largely depends on the size and number of soil pores and on the amount of water logging.

Agriculture: An activity of man which is primarily aimed at the production of food, fiber, fuel etc by optimum use of terrestrial resources.

Agrochemical: Biologically active chemicals used in agriculture which includes insecticide, fungicides, herbicides, growth substances etc.

Air storage: Storage in the room at normal air temperature providing proper air flow and ventilation which facilitates the removal of ethylene produced by fresh fruit and vegetables and maintain the CO_2 and O_2 level about 0.03 and 21.0 percent respectively.

Alkaline soil: A soil having more than 15 percent exchangeable sodium ions and alkaline in reaction throughout the root zone and precisely any soil having a pH value greater than 7.0.

Alluvial soil: A soil developed from recently deposited alluvium with no horizon development or modification of the recently deposited materials.

Ammonia injury: Refrigerated fruits and vegetables may get damaged by escaped ammonia gas (Refrigerant) which render brown to greenish black discolouration of outer the tissues and also softening of the tissues.

Ammonification: Conversion of organic nitrogen to ammonium ions (NH^+_4) by the micro organism present in soil.

Ammonium fixation: The adsorption or absorption of ammonium ions by the mineral or organic fractions of the soil in such a manner that they are relatively insoluble in water and relatively non exchangeable by the usual methods of cation exchange.

Analytic seed sample: The seed sample obtained by diving the mechanical seed sample from which the analysis for different tests are carried out conveniently.

Anion: Negatively charged ion, ion which during electrolysis is attracted to the anode.

Anion exchange capacity: The sum total of exchangeable anions that a oil can adsorb. Expressed as milliequivalents per 100 grams of soil.

Arable crops: Crops which require cultivation.

Arable land: Cultivated land used for growing crops.

Alternate: Arrangement of a bud or leaf that occur singly at node.

Annual: Plant in which the entire life cycle its completed in a single growing season.

Anther: Upper portion of a stamen, containing the pollen grains.

Anthesis: An opening of a flower bud, the period of pollen distribution.

Aphid: Very small insect commonly called plant lice.

Balanced diet: It may be defined as one which contains different types of foods in such quantities and proportion that the need for calories, amino acids, vitamins, minerals, fats and carbohydrates are adequately met for maintaining health, vitality and general well being.

Band placement: Application of fertilizers in bands to one or both sides of the seed or plant.

Band seeding: Placing the seed in rows directly above but not in contact with a band of fertilizers.

Barren: Unproductive in terms of fruit or seed set.

Basal plate: The solid tissue at the bottom of a bulb to which the fleshy scales are attached.

Base temperature: The threshold temperature level below which plant will not develop and each plant has its own base temperature. *e.g.* pea (4.4°C), French bean (10°C), Asparagus (5.5°C), Spinach (2°C), Pumpkin (13°C), tomato (15°C) etc.

Basin irrigation: A surface irrigation method mainly for field crop in which flat interconnected basins are made around the plant trunk and irrigation water is let in through the connecting channels to the basin only.

Bed: Ridge of soil formed with furrows located on each side in which row crops are sown or transplants are planted for planting later in the field.

Beer: A class of alcoholic beverage brewed from malt or malt substitute with the addition of hops to give a bitter taste.

Bench terrace: A series of step like, mostly leveled platforms cut into the hill slope for rendering cultivation in the sloping land.

Betalains: Red and yellow pigments found in garden beet which can be hydrolyzed into a sugar and a coloured portion.

Bhasinda: The underground stem of *Nelumbium* a popular vegetable in Kashmir.

Biennial: Plants having a two year life cycle –vegetative in the first season and reproductive in the second season and this transition from vegetative to reproductive stage is often required environmental trigger such as vernalization or photoperiod. *e.g.*, cabbage, onion, carrot etc.

Bin: An enclosed structure used for storage of seeds.

Bin drier: A drier used to complete drying of foods after removing most of the moisture in a tunnel drier or its equivalent.

Biological yield: Total dry matter yield of a crop in a unit land area.

Biomass: The total amount of living matters in a given area, expressed in terms of living or dry weight per unit area.

Biopesticide: Preparation or formulations manufactured to be used in the control or eradication of disease, pest, or weeds in which the active ingredient or principles is based on living micro-organism, or is derived without significant purification or modification from bacteria, fungi, nematodes and protozoa.

Biosynthesis: Formation of complex compound from simple substances by living organisms.

Bitterness in carrot: A storage disorder due to deleterious effect of ethylene to carrot roots which causes an increasing in the total phenol content of roots and induces the formation of new compounds including isocoumarin and eugenin and isocumarin is mainly associated with bitter flavor in carrots.

Blanching: Heat treatment of vegetables in boiling water or steam for 2-5 minutes prior to canning for inactivation of enzymes, removal of air, setting of natural

Blanket spraying: Spraying of herbicide, insecticide etc uniformly over the entire area.

Blind cultivation: Cultivation before the planted crop emerges.

Bolter: Sporadic occurrence of abnormally big sized tuber in potato.

Border crop: Crops which are grown around the field boundaries in narrow strips with twin objectives of protecting the main crop from insect pest and stray cattle and producing livestock feed.

Border rows: The recommended number of rows of the male parental lines grown on all the sides of hybrid seed field.

Border strip irrigation: A surface irrigation method where the field is divided into a number of long narrow strips with small parallel ridges on the sides and individual strips after perfect leveling are connected with the water supply channels.

Breeders seed: Seed or vegetative propagating material directly controlled by the originating or in certain cases the sponsoring breeder or institution which is provided for the initial and recurring increase of foundation seed. This category of seed is genetically cent percent pure.

Breeding cycle: The seed-to-seed cycle.

Breeding system: A particular mating system that involves a certain type of plant material together with the necessary selection procedures.

Broadcast: Scattering seeds or fertilizers uniformly over the soil surface rather than placing in rows.

Caducous: Non-persistent conditions of different plant parts which are shed within the life cycle of the plant *e.g.* sepals falling off as flower opens in okra, poppy etc. or stipules falling off as leaves unfold in lime.

Calcareous soil: A soil containing sufficient calcium carbonate (often magnesium carbonate) and alkaline in reaction.

Calcifuge: Plants that grows best on acid soils.

Canning: The process of storing fruit pieces in 35 -50 percent sugar solution and vegetable pieces after blanching in 2-3 percent salt solution together with required citric acid or ascorbic acid in the can.

Capillary water: Water that is retained around the soil particles and in the capillary pores of the soil.

Capsaicin: Crystaline, colourless, pungent principles ($C_{18}H_{27}O_3N$) of chilli which is the condensation prouct of 3-hydroxy-4 methoxy benzylamine and decylenic acid, secreted by the outerwall of the fruit.

Capsanthin: A carotenoid pigment responsible for the characteristic orange-red coloration of ripe chilli.

Cardinal temperature: Plant growth cases below a certain minimum and beyond maximum temperatures and there is an optimum temperature between these limits at which growth proceeds with greatest rapidly. These three points are known as cardinal temperatures.

Carotene: The orange red carotenoid pigment which is the precursor of vitamin A. one molecule of beta carotene is converted into two molecules of Vitamin A by hydrolysis.

Carotenoids: A group of yellow, orange and orange red fat soluble pigments present in plnat parts. These are either hydrocarbons or its derivatives and are composed of isoprene units. Carotenoids occur in different forms like carotene, lycopene, lutein, violaxanthin, neoxanthin, capsanthin, bixin, xanthophylls etc.

Cash crop; A high value marketable crop *e.g.* vegetables.

Catch crop: A crop grown temporarily in order to get additional productivity out of the land prior to the establishment of the main crop.

Cation: Positively charged ion, ion which during electrolysis is attracted to the cathode.

Cation exchange capacity: The sum total of exchangeable cations that a soil can absorb. expressed in miliequivalents per 100 grams of soil.

Celery lettuce: Stem type cultivar of lettuce, grown for its thick stem which is eaten after peeling.

Centers of diversity: Areas where cultivated plant species and or their wild relatives show much greater variation than anywhere in the rest of the world.

Centre of origin: An area where the given plant species is believed to have originated releasing its greatest genetic diversity in that area.

Certified seed: It is the progeny of foundation or registered seed that is so handled as to maintain satisfactory genetic identity and purity and that has been approved and certified by the certifying.

Check irrigation: Irrigation method where ridges are made around the area to be irrigated to retain water there.

Check row planting: The process of planting in which row to row and plant to plant distances are uniform and plants across the rows are also in line.

Chemical dormancy: Type of seed coat dormancy in which germination inhibiting chemicals *viz*, various phenols, coumarin and absicisic acid accumulated in the fruit as well as in the seed covering strongly inhibit germination, found in cucurbits, tomato etc.

Chemical preservation: Food additives which are specifically added to prevent the deterioration or decomposition of food by microorganisms, food enzymes or by purely chemical reactions. Thses include sodium benzoate, potassium metabisulphite, boric acid, propeonic acid, lactic acid, etc. But do not include salt, sugar, antioxidant etc.

Chilling requirement: Cold period required for certain plants to break physiological dormancy or est and it is expressed in terms of the required number of hours at 7°C or less.

Chilling storage: Storage at temperatures not far above freezing which usually involves cooling by ice or by mechanical refrigeration.

Chlorosis: A disorder showing yellowing between the veins on upper leaf surface due to loss of chlorophyll.

Class: A group that includes varieties of similar magnitude.

Clarification: The process by which a liquid can be made into clear and transparent by sedimentating or removing the suspended materials present in it.

Classification: The systematic methods of arrangement of plants in categories according to some definite sequences, categories in common use are Division, Class, Order, Family, Genus, Species and Variety.

Clay: Soil material containing more than 40 percent clay, less than 45 percent sand and less than 40 percent silt.

Clay loam: Very fine textured, poorly drained, fertile, slightly acidic or alkaline soil.

Clean cultivation: Periodic soil tillage to eliminate all vegetation other than the crop being grown.

Climate: A long term prevalent weather conditions of an area, determined by latitude, altitude etc.

Climatic index: A number which condenses climatic data into a simplified expression.

Cloves: Bulb lets present in the bulb of garlic which are produced from the axillary buds of the younger foliage leaves.

Clump: Aggregate stems arising from the same root rhizome as in turmeric, ginger etc. or from stool.

Coastal alluvium: Soils that are subjected to the action of sea water, saline, sandy loam in texture, excessive in drainage, low in water holding capacity and poor in plant nutrients.

Coated fertilizer: fertilizer granules with thin covering of a different material in order to modify the characteristics of the fertilizer.

Cold frame: An enclosed (glass or plastic), unheated frame used as a propagation unit and for growing or protecting young plants in early spring in a temperate climate.

Cold hardiness: Acquired ability of a plant to a withstand cold condition due to continuous exposure to low temperature.

Cold storage: An insulated storage with refrigeration to maintain a stable cold temperature for long term storage of perishable products.

Companion crop: Crop grown with another to secure an earlier or larger return than from one crop alone.

Complex fertilizer: The commercial fertilizers containing at least two or more of the primary essential nutrients.

Compost: Bulk organic wastes which are decomposed under controlled conditions in piles, pits or bins used as manure.

Composting: A biological process in which microorganisms decompose the organic matter and lower the carbon-nitrogen ratio of the refuse and the final product of composting is a well rotten manure known as compost.

Conical root: when the root is broad at the base and gradually tappers towards the apex like a cone as in carrot.

Conservation cropping: farming that aims at providing protection against soil erosion for sub-stained farm productivity.

Contour border irrigation: Border strip irrigation practiced along the contour.

Contour farming: Cultivation along the contour slope to reduce run off, conserve moisture and increase crop yield.

Controlled atmospheric storage: Storage of commodities under gaseous atmosphere, like gas storage, where the composition of atmosphere is controlled accurately. The modified atmosphere and controlled atmosphere differ only in degree of control and CA is more exact. This storage method in combination with refrigeration markedly enhances storage life of fruits and vegetables.

Conventional tillage: Primary deep tillage operation followed by secondary tillage for seed bed preparation.

Cool storage: Storage at a temperature range from -2°C to 15°C depending on the food.

Cost of cultivation: Total expenditure involved in raising a crop including rental value of the land.

Cover cropping: Intercropping mainly on slopy and terraced land to establish a vegetative cover for the protection of soil from erosion.

Cover crops: Crops that are grown both for the protection of the soil from erosion and for soil improvement *e.g.* cowpea.

Critical period: The period during which a crop is affected severely due to moisture stress and the loss cannot be compensated by adequate moisture supply in other periods or stages.

Cropping index: The number of crops grown per annum on a given area of land which is expressed in percentage.

Cropping intensity: Ratio between total cropped area and actual net cultivated area without considering the length of growing periods for various crops, and is expressed in percentage.

Cropping pattern: Yearly sequences and spatial arrangements of crops or crop and fallow on a given area.

Cropping scheme: The plan of raising crops on a farm with an object of getting the maximum returns from each crop without impairing fertility of the soil.

Crop production: Exploitation of morphological and physiological responses of the crop within a soil and atmospheric environment to produce a high yield per unit area of land.

Crop Productivity: Economic yield of a crop per unit area.

Curing: A post harvest treatment of tuber and bulb crops by exposing them to relatively high temperature and high humidity to facilitate the drying of upper skin and suberization of the outer tissue, periderm formation and healing of injured surface, reduces the moisture loss through transpiration prevents microbial attack, reduces rotting and resulting more shelf-life.

Curing (Preservation): A pre-treatment of the raw materials before being processed (mainly for pickle preparation), done by treating the materials with dry salt or brine (8-12 per cent) for a sufficient time to facilitate processing and to control the microorganisms and fermentation of the materials. During curing the vegetables lose their raw flavor and become good textured, firm and crisp.

Damping off: Death of small seedlings due to attack of certain fungi like *Pythium, Rhizoctonia* etc.

Day neutral plant: Plant in which flowering is not influenced by day length *e.g.* tomato, cucumber, okra, asparagus, capsicum snap bean, etc.

Deep soil: A soil having a solum depth of more than 100 cm.

Dehaulming: Removal of the top portion (Haulm) of potato in the seed crop to avoid the infestation of virus carrying insect vectors.

Dehydration: A drying process where moisture of foods is removed by artificially produced heat under controlled conditions of temperature, relative humidity and air flow.

Denitrification: Biochemical reduction of nitrates to nitrous oxide (N_2O) and then to free N_2 which takes place in soil in the presence of active microbial (*Nitrosomonas* sp.) population.

Density: Mass per unit volume.

Desalinization: Removal of salts from saline soils usually by leaching.

Dibbling: Sowing or planting propagation materials in holes made with a planting peg.

Direct sowing: Sowing seed directly on an area where the crop is to be raised.

Diurnal: Flower opening during the day and closing at night.

Diversified cropping: A cropping plan in which no single crop contributes 50 per cent or more towards the total crop production or monetary income (comparable equivalents) annually.

Dormancy: Alive condition but in rest with reduced metabolism of seed, bud, etc.

Double cropping: Taking two crops in a year.

Drainage: removal of excess surface or ground water from the land by means of surface or sub surface drains.

Drip irrigation: A subsurface irrigation system where water is applied slowly drop by drop to the root zone of the crop through drip type nozzles.

Drought: The moisture deficit condition in the soil when the demands of potential evapo-transpiration are not fulfilled.

Dry land farming: The practice of crop production in low rainfall areas without irrigation.

Dry matter content: Plant substances with minimum moisture content, accomplished by oven drying usually at 70°C for 72 hours.

Dry storage: Storage without extremes of temperature change at above 20°C and below 50 percent relatively humidity.

Economic yield: Economically useful part of the total dry matter produced by a crop in a unit land area.

Economic threshold yield: Yield per hectare required t meet the total expenditure each year.

EMS: A chemical mutagen, chemically Ethylmethane sulphonate.

Endospermic seed: The seed in which the embryo is reduced in size in proportion to the rest of the seed, and in these seeds the storage material lies in the endosperm and perisperm.

Eradication: A method of disease control in which the pathogen is eliminated after it has been established.

Experimental yield: the yield level of a crop/variety obtained at experimental stations where yield maximization is the major objective using all possible technologies.

Fallow land: Crop land left uncultivated for one or more seasons in order to restore productivity mainly through accumulation of water, nutrients or both.

Feeder roots: Fine roots and rootlets with a large absorbing area (root hairs).

Fertigation: Application of fertilizers with irrigation water. its applicability depends besides the method of irrigation used, on the type of soil and crop, the climatic conditions or water quality and the types of fertilizers available.

Fertility: Ability of an organism to produce viable progeny.

Fertility gradient: The variation in natural soil fertility in any direction across an area.

Garden for vegetable processing: Type of vegetable farming where vegetables are produced with a sole objectives of supplying them to the processing factories.

Gas storage: Storage of produces under modified or controlled atmosphere where atmospheric compositing surrounding the commodity is different from that of normal air (78.08 per cent N_2, 20.95 per cent O_2, 0.003 per cent CO_2) which usually involves reduction of oxygen and or elevation of carbon dioxide concentration. This storage reduce ripening, senescence and respiration and threreby increase shelf life of the produce.

Gelmeter: A long pipette like glass apparatus used for determining the concentration of pectin present in fruit extract. It gives direct reading or indication of the amount of sugar needed for one litre of extract.

Geotropism: Movement of plant organs in response to the force of gravity.

Germination: The process of reactiviation of the metabolic machinery of the seed following the emergence of the radicle and plumule leading the production of a seedlings.

Germination percentage: The percentage of the pure seed of the kind under consideration which produce normal seedlings.

Girdling: Process of constricting the periphery of a stem which blocks the downward translocation of carbohydrates, hormones, etc. beyond the constriction which rather accumulates above it.

Glabrous: Smooth surface and free from hairs or outgrowths of any kind.

Glass house: A structure of glass with or without inside heating arrangements where plants are either grown or conserved doing difficult seasons.

Glaucous: Covering the leaf or stem surface with waxy coating which causes a shiny and either whitish or bluish tinge as seen in *Nymphae* sp.

Gluconeagenesis: The process by which phosphoesolpyrurate (PEP) generated from pyruote is converted by a series of enzyme mediated reaction to glucose-6-phosphate.

Glycolipid: A lipid with a sugar esterified to the third hydroxyl group instead of phosphate. It constitutes about 80 per cent of the lipid fraction of chloroplast in higher plant.

Glycolysis: A series of biochemical reactions in respiratory process in which hexose sugar is converted to pyruric acid with net gains of two ATP molecules.

Grading: The process of sorting products into different lots conforming to certain predetermined standards.

Graft hybrid: Origination of new plants from the fusion of nuclei of vegetative cells of two different individuals which is comparable to trace somatic hybridization and grafting may induce fusion of cells.

Graft incompatibility: Inability of the stock and scion to make a successful graft union.

Grafting: Art of joining two separate plant parts together so that by tissue regeneration they form a union and grow as one plant.

Gram calorie: The amount of heat required to raise 1 g of water from 0°C to 1°C.

Gvartational water: Soil water in excess of the capillary capacity which is incapable of being retained under conditions of free drainage.

Greenhouse: A structure in which plants are either conserved or grown in a difficult season.

Grass return: Total income from the farm by virtue of sales of entire farm produce.

Ground cover: A mass of dense low herbaceous plants and shrubs that grow over the surface of the soil mainly to check soil erosion.

Growth medium: Any material such as soil, peat etc. used as a support for plant roots, that has a capacity for water retention and which may contain added and/or naturally occurring nutrients.

Habit: Shape, generally appearance and mode of growth common to the individuals a species or other systematic group.

Hakuran: An artificial amphidiploid of cabbage on Chinese cabbage produced through embryo culture technique which is a good leafy vegetable.

Half-hardy vegetable: Vegetable crops which can thrive well in cool weather condition but cannot tolerate frost *e.g.*, beat, carrot, cauliflower, lettuce etc.

Half-slip: A harvesting index of muskmelon or cantaloupes for distant market when netting is not complete and fruit can be separated from stem by twisting.

Hardening: Subjecting plants to adverse conditions to haster Tissue maturation for increasing hardness, mainly done by with holding irrigation.

Hardy vegetable: Vegetable crops which are tolerant to frost *e.g.* asparagus, broccoli, cabbage, peas, turnip, garlic, knol-khol, radish, spinach, onion etc.

Harvest: A single deliberate action to separate a crop from its growth medium (A plant part from the plant or the whole plant from soil).

Harvest index: A productive parameters depicting the economic productivity interms of dry matter and is calculated as $\dfrac{\text{Economic yield}}{\text{Biological yield}} \times 100$.

Harvest maturity: Attainment of a particular maturity index by a plant part of the plant as whole after which it is harvested.

Heading efficiency ratio: A head character of cabbage, Chinese cabbage, calculated by dividing mean head weight (g) with mean non-wrapper leaf weight (g).

Head shape index: A head character of cabbage, Chinese cabbage, calculated by dividing mean head length (cm) with mean head width (cm).

Head to seed method: A method of seed production practiced in cabbage where, the selected plants with fully matured heads are lifted prior to snowfall, stored and again replanted a the onset of spring for seed production.

Heat injury: A disorder due to exposure of the tissues to direct sunlight or to excessive high temperature and symptoms of it include bleaching of colour, surface burning or scalding, uneven ripening, excessive softening, desiccation of tissue etc.

Heat tolerance: Ability of the plant to grow under high temperature conditions which is a genetically controlled attribute.

Heavy soil: A fine textured soil like clay or clay loan characterized by low infiltration capacity, poor drainage and inadequate aeration.

Hectare: Area measure in the metric system, equal to 10,000 square meters or 2.471 acres.

Hilum: The point of attachment of the seed to its stalk.

Horizontal resistance: Resistance which is pathotype, non-specific and incomplete but permanent and the inheritance is polygenic. It is also called field resistance.

Horticultural maturity: The stage of development of a plant part which possesses the necessary pre-requisites for utilization by consumers for a particular purpose and it depends on the particular crops are harvested at fully developed stage but okra, beans etc. are harvested at tender.

Humid areas: Crop production area which receive sufficient supplies of rain annually and require the application of irrigation as a form of insurance for the production of high yields.

Hygroscopic water: Portion of the soil water which is retained as a thin film by the soil particles and is associated mostly with the colloidal part of the soil.

Hypogenous germination: A pattern of germination where the lengthening of the hypocotyl does not raise the cotyledons above the ground and only the epicotyl emerges as seen in pea.

Hypsometer: An instrument for determining the height of standing tree from some distance.

Immunity: Absolute resistance of a plant against the adverse effects of the micro-organisms, insect, pest etc.

Impermeable seed: Seed whose seed coat allows no passage through to water.

Impervious soil: Soil through which air, water or plant roots penetrate very slowly, if at all.

Implements: Equipments used to carry out different form operations.

Improved production technology: Use of improved varieties and associated improved crop management practices followed on a form.

Improved seed: Genetically and physically pure seed of as improved crop variety. Different categories of improved seeds are nucleus seed, breeder seed foundation seeed, registered seed and certified seed.

Indeterminate type: A growth habit where, the main axis of the plant continues to grow indefinitely *i.e.,* emergence of inflorescence and elongation of stem go on side by side.

Indigenous: Native to a specified region.

Induced dormancy: Seed dormancy which is induced by such exernal factors as unfavourable water supply or temperature through the embryo is quiescent.

Infusion: A methods of seed priming where the seeds are immerse in acetone and dischloromethane solution for 1 to 4 hours for the chemicals to be infused into the seeds following removal of the solvent by evaporation. The incorporated chemical to observed directly into embryo with subsequent soaking in water.

In situ: Refers to natural or original position.

Integrated pest management: A pest management practice which involves a mix of practices such as use of resistant varieties, managing the natural predators of pests, cultural practices and judicious application of pesticides to control the pests.

Interaction: The extent to which the effect of one factor varies with changes in the strength, grade or level of other factors in an experiment.

Interculture: Cultivation practices performed in the standing crop.

Internal dormancy: Seed dormancy which is internally controlled within the living tissues of the seed, also called inate dormancy.

In vitro: A Latin word meaning 'in glass' which implies that biological process are made to occur outside the body of the organism in an artificial environment like in glass.

In vivo: A Latin word meaning 'in living' which implies that biological process are carried out in the living organisms.

Irrigation: Application of water to soil for the purpose of supplying moisture essential for plant growth.

Irrigation efficiency: The ratio of water actually consumed by crops on an irrigated area to the amount of water diverted from the source onto the area.

Isolation distance: A specified distance among the population for prevention of intercrossing in order to maintain the genetic purity of the seed and it depends on the breeding behaviour of the crop and category of the seeds to be produced.

Jelmeter: An aparatus used for pectin test during the preparation of jam, jelly. It directly indicates the amount of sugar to be added per litre of extract.

Juvenile stage: Early or vegetative phase of growth characterized by carbohydrate utilization and before which flowering cannot be induced in plant.

Ketchup: A concentrated juice or pulp without seed and skin where spices, salt, sugar, vinegar etc. are added to the extent that it contains not less than 12 per cent vegetables or fruit solids and 28 per cent total solids.

Kitchen gardening: Growing of vegetable crops in the residential houses to meet the requirements of the family all the year round.

Land: Part of surface of lithosphere not usually covered with water which forms the natural and cultural environment where crop production takes place.

Land equivalent ratio: It denotes the relative land area under sole crop required to produce the same yield as obtained under a mixed or intercropping system at the same management level.

Land levelling: The reshaping of the land surface to facilitate a more uniform application of irrigation water or for preventing soil erosion.

Land reclamation: Making land capable of more intensive use by changing its character and/or environment through such operation as land clearing, controlling erosion, construction of irrigation facility etc.

Leaching: Process of removing soluble materials through water from soil, seed etc.

Leaf area density: It is the ratio between leaf area index (LAI) and plant height and expressed as cm^2/m.

Leaf area duration: A measure of the ability of the plant to produce and maintain leaf area which is obtained by integrating the leaf area index over crop growth period.

Leaf area index: The ratio of leaf area (one surface only) of a crop to the ground area on which it grows.

Leaf mulch: Intercropping is referred to as life much because it conserved soil and reduces evaporation loss from the interspaces.

Light soil: A coarse textured soil like sand or sandy ioam characterized by high infiltration and good drainage capacity.

Light transmission ratio: Ratio of light intensity at he ground level of a population (LG) and light intensity of the top of the population (LP).

Liming: The process of soil amendment to neutralize the soil acidity by addition of line in the soil.

Liquid fertilizers: Commercial fertilizers in liquid from such fertilizers are chiefly anhydrous ammonia, aqueous solution of nitrogen and some mixed fertilizers which are applied to the soil through irrigation water, starter solution etc.

Liquid manure: Liquid from animal urine, litter juice or from a dung heap.

Loam: A textural class name for soil having a moderate amount of sand, silt and clay and well supplied with organic matter.

Local variety: Unimproved plant types that are well adapted to a particular locality/environment.

Long day plant: The plant which requires a day longer than its critical day length for flowering *e.g.*, lettuce, radish, onion, cabbage, carrot etc.

Low acid foods: Foods having pH 5.0 and higher which are liable to spoilage by thermophiles and mesophilic putrefactive anaerobes including *Clostridium botulinum e.g.*, peas, beans, asparagus etc.

Lycopene: Red pigment found in ripe tomato which is a straight chain derivative of carotene with no vitamin activity. Its chemical composition ($C_{40}H_{56}$) is same as that of carotene.

Lysimeter: A device for measuring percolation and leaching losses from a column of soil under controlled conditions.

Macro element: Nutrient elements that are required in relatively large quantities by the plants *e.g.*, N, P, K, Ca, Mg and S.

Makhana: The seeds of *Euryele ferox* (A prickly aquatic plant resembling lotus) when parched or roasted over hot sand give a puffed up product, commonly called makhana.

Market gardening: Vegetable farming for supply of vegetables to the consumers in the local market; one of the most intensive types of vegetable farming.

Maturity index: The factors for determining the harvesting of fruits and vegetables according to consumer's purpose, types of commodity, etc. and can be judged by the (a) visual means (colour, size, shape etc.) (b) physical means (firmness, softness etc.) (c) computation (heat unit) and (d) physiological methods (respiration).

Mean square: The variance so called because it is the average of the squared deviations from the arithmetic mean of a series of values.

Median: Middle value in a given set of observation.

Medium acid foods: Foods having pH 5.0-4.5 which are usually spoiled by thermophilic anaerobes *e.g.* vegetable mixture, soaps, sauces etc.

Mericloning: Vegetative multiplication through meristem culture.

Meristem culture: A type of *in vitro* culture technique where shoot or apical meristem along with some surrounding tissue in grown on suitable culture media which is employed for clonal propagation and recovery of virus free plants.

Messanger RNA (mRNA): Ribonucleic acid (RNA) capable of carrying information for the amino acid sequences of specific proteins. The information is codded in chromosomal and DNA. It moves from the nucleus to the ribosomes where the massage for specific protein synthesis is deciphered with the help of ribosomes and specific RNA's.

Metaeryotic liquid: During storage, the unfrozen concentrated solutions of sugars, salts etc. may ooze out from package of fruits or concentrated as a viscous materials called metaeryotic liquid.

Metsubre: A nutritional disorder of taro (*Colacasia esculenta*) due to calcium deficiency where the defective corm have smooth or concave top, slightly brownish in colour and are of varying size.

Microclimate: Local climate conditions near the ground or area around plants resulting from the modifications of the general climatic conditions by local difference in relief, exposure and cover etc.

Micron: One millionth part of a meter.

Micronutrients: The essential plant nutrients that are required in small quantities *e.g.*, iron, manganese, boron, molybdenum, copper, zinc, chlorine etc.

Micropyle: A small slit or opening near the hilum of seed.

Mineralization: The conversion of an element from an organic form to an inorganic state as a result of microbial decomposition.

Minimum description: Minimum number of distinct morpho physiological features that can effectively discriminate among genotypes for evaluation of germplasm collections.

Minisett: Seed tuber pieces of yam (*Dioscorea alata*) which are used for rapid multiplication.

Mixed intercropping: Growing two or more crops simultaneously with no distinct row arrangement.

MMS: Methylmethane sulphonate, a chemical mutagen.

Morphological dormancy: A type of seed dormancy in which the embryo is not fully developed at the time of ripening and such seed may have either rudimentary or undeveloped embryo found in carrot.

Mucilage: A complex glutinous carbohydrate secreted by certain plants (Okra, Basella etc.) which absorbs water freely.

Mulch: Any material such as straw, plant residues, leaves or plastic film placed on the soil surface to reduced evaporation, erosion or to protect plant roots from extremely low or high temperatures.

Mulch film: A low density polythylene film that protects crops against thrips, aphids, and other photophobic pests by reflecting ultraviolet rays as these pests tend to gather around strong visible rays but stays away from ultraviolet rays.

Mulching: Practice of covering the soil surface for conservation of soil moisture and control of weeds and soil erosion. It may add organic matter to the soil where plant wastes are used as mulch.

Multiple correlation coefficient: A coefficient that measures the degree to which the dependent variable is influenced by a series of other factors studied.

Multiple cropping: Growing of two or more crops simultaneously where there is significant amount of inter crop competition.

Multiple green house: Greenhouse which are built side by side using a common wall between the two structures where different micro-climatic conditions are maintained in different houses in order to suit the requirement of different crops.

Multivariate analysis: An useful statistical tool for quantitative estimation of genetic diversity.

Mutagen: A substance or treatment which under suitable conditions may cause mutation.

Mutagenesis: Induction of mutation by treatment with mutagens.

Mutation: A sudden new variation in the genetic constitution due to individual gene or chromosomal changes.

Naked chilling: Stratification of seed in plastic bags without a surrounding medium.

Napiform root: Root when swollen becoming spherical at the upper part and sharply tapering at the lower part, as in beet, turnip etc.

Nastic movement: A response of the plant parts that is independent of the direction of the external stimulus such as opening of buds caused by an alteration in light intensity.

Naturalization: The process of establishment of a plant brought into a region where the conditions are similar to those whence it came.

Natural selection: An important feature of certain theories of evolution and according to which, agents other than man determine which member of a population will survive. It refer of differential rate of reproduction of different genotypes of an organism in response to environmental factors.

Necrosis: Death of plant tissues due to disease and frost.

Nectar: Sweet liquid secreted by the nectaries of flowers which often attract pollinators.

Nematicide: Chemicals used to destroy nematodes or ell worms *e.g.* CS_2, Paradichlorobenzene, $(C_6H_4Cl_2)$ or PDB, EDB, DBCP, methyl isothiocynate etc.

Nematode: Minute thread like ellworms which cause injury to the plants.

Net assimilation rate (NAR): It is the dry weight accumulated per unit of leaf area per unit time which expresses the photosynthetic efficiency of a plant.

Net productivity: The arithmetic difference between calories produced in photosynthesis and calories lost in respiration.

Net return: Income obtained or remaining after deducting cost of cultivation from the grow return.

Niche: The role of an organism in the environment, its activities and relationships to the biotic environment.

Nick: The two parents for hybrid seed production are said to "nick" when they produce high yields of seed of a highly productive and desirable hybrid.

Nitrification: Production of nitrate (NO_3^-) from ammonium ion (NH_4^+) in the soil which is a two-stage biological process: Oxidation of NH_4^+ and production nitrite (NO_2^-) via hydroxylamine by Nitrosomonas and related genera of bacteria and then oxidation NO_2^- to NO_3^- by Nitrobacter bacteria.

Nitrogen fixation: The process by which atmospheric N_2 is reduced to ammonium ion (NH_4^+) principally by free living soil bacteria, blue green algae or bacteria associated symbiotically with roots, especially of legumes.

Nodulation: Development of knob like outgrowth in the roots of leguminous plants in response to the stimuli of *Rhizobium* bacteria which is involved in fixation of atmosphere.

Non-climacteric fruits: Fruits showing a gradual decline in respiration rate with ripening. These fruits on the tree and therefore they may be harvested when they become edible *e.g.*, grape, cucumber, cashew, ber etc.

Non-selective herbicide: A chemical that is toxic to plants generally without selectivity to species.

Non-symbiotic nitrogen fixation: Fixation of molecular nitrogen by free-living micro-organisms like *Clostridium, Azotobacter, Azospirillium* etc.

Normal distribution: A theoretical distribution based on continuous variable. This distribution is represented by a bell shaped curve *i.e.*, its frequency curve is symmetrical about mean and the distribution in unimodel.

Normal solution: Gram equivalent weight of a compound when dissolved in one litre of water the solution is one normal solution.

Nursery: An area where the planting materials are raised for sowing or planting in gardens or fields. In other words, the nursery industry involves the production and distribution of different kinds of planting materials.

Nursery bed: A prepared area where seed is sown or into which transplants or cuttings are planted.

Nutrient: Elements available from soil, air and water which are utilized by the plant for growth, development and reproduction.

Nutrient recovery percentage: Amount of fertilizer utilized by the crop.

Nutrient use efficiency: Characteristic of a genotype to more efficiently use the extracted mineral nutrients in plant growth and yield. Breeding of more nutrient efficient type in very important to prevent excess fertilizer application.

Nutrition: The sum of the process by which plant absorbs and utilize food substance.

Nutritive value: It is characterized by the chemical composition, digestibility and nature of the digested products of the food materials.

Nyctophilic plants: Syn. Short day plant.

Nyctophobic plants: Syn. Log day plant.

Off-type: Any type of plant which is outside the acceptable limits of varietal variation.

Oleoresin: A natural combination of resinous substances and essential oils present in certain crops like chilli.

Olericulture: A Latin term used to designate science and management of vegetable crops.

Order: A taxonomic group which contains one or more families.

Organic soil: A soil which contains a high percentage (>20 per cent) of organic matter throughout the solum.

Organoleptic evaluation: Descriptive index characters of the fruits or cooked samples of vegetables for appearance, firmness, taste, mouthfeel etc.

Orthodex seed: A category of seeds which can be dried to a low moisture level and would show a loss of viability with the rise in moisture content and are usually long lived *e.g.* most vegetable seeds.

Osmoconditioning: A method of seed priming where seeds are placed in a shallow layer in a container of 20-30 per cent polyethylene glycol solution, which may include hormones or fungicide and after incubating at 15°C to 20°C for 7-21 days the seeds are washed with distilled water, air dried at 25°C, stored and then planted directly in their permanent location.

Overhead irrigation: Syn. Sprinkler irrigation.

Oxylophytes: Plants tolerant to high acidic soil conditions.

Paired t-test: This t-test is applied to test the significance of mean difference between paired observations in the normal population of the same environment.

Palynology: Study of pollen grains and other parts of flower belonging to the past geological ages.

Panicle: Main axis of the raceme which is branched and the lateral branches bear the flower as in Delonix.

Papilla: Small, fleshy protuberance on the surface of the leaf or flowers.

Parameters: A numerical quantity which describes a particular characteristic of a population.

Parasexual hybridization: Syn. Somatic hybridization.

Parietal placentation: When placentae bearing the ovules develop on the inner wall of the one-chambered ovary corresponding due to the confluent margins of carpels as seen in radish.

Parthenocarpy: Formation of seedless fruit from ovary without fertilization.

Partial correlation coefficient: The correlation between two series of variables independent of the accompanying variation due to other variables.

Parts per million (ppm): Quantity of a substance contained in a million parts of a mixture or solution.

Passive absorption of water: Absorption of water by the roots as a result of forces originating in the leaves due to transpiration pull.

Path coefficient: Statistical analysis which provides an effective means of entangling direct and indirect causes of association and at the same time measures the relative importance of each casual factors.

Peat soil: Unconsolidated soil material consisting primarily of undecomposed or slightly decomposed organic matter which accumulate under conditions of excessive moisture.

Pedicel: A stalk on which an individual blossom is borne.

Pelleted seed: Seeds coated with an adhesive material such as clay, activated char coal etc. which making them uniformly round for precision planting.

Pencil stripe of celery: A disorder associated with high soil phosphorus showing narrow brown lines on petioles.

Pepo: Fleshy, many seeded fruits develop from an inferior, one celled or spuriously three celled, sycarpous pistil with parietal placentation *e.g.* cucumber, melons, squash, gourds etc.

Percolation: Downward movement of water through the soil under gravity or hydrodynamic pressure or both under saturated or nearly saturated conditions.

Perennial: The plants which live for more than two years and repeat the vegetative-reproductive cycle annually.

Perfect flower: Syn. Hermaphrodite flower.

Perianth: When the calyx and corolla do not differ much in shape and colour, they together are said to form the perianth as seen in onion, garlic etc.

Pericarp: Part of a fruit, developed from the wall of the ovary which may consists of epicarp, mesocarp and endocarp.

Perishable crops: The crops that can not be stored for longer period at room temperature such as fruits, vegetables etc.

Perse performance: Quantified response of a crops stand.

Petting: Removal of seedlings from a seed bed to a pot.

pH: A measure of acidity or alkalinity which is expressed as the negative log of the hydrogen ion concentration; pH of 7.0 is neutral, less than 7.0 is acidic and more than 7.0 is alkaline.

Phendices: Compounds with benzene rings and various attached substituent groups such as hydroxyl carboxyl or methoxyl and other non-caronatic or ring structures. Phenolics include aromatic amino acids, simple phenols, polyphenols and derivatives like phytoalexins, coumarins, lignin, anthocyanin, flavonols and flavones.

Phospholipid: A lipid with a phosphate group esterified to the third hydroxyl group of glycerol and often other ions are esterified to the phosphate *e.g.* choline, glycerol, inositol, ethanolamine.

Photodormancy: A type of physiological dormancy of seed where germination of seed is sensitive to light *i.e.* seeds of some plants require light to germinate, whereas others require darkness *e.g.* lettuce seed require light and Allium, Amaranthus etc. seeds require darkness for germination.

Photoperiodism: The reaction of various plants to the daily fluctuations in radiant energy.

Photosynthesis: Biochemical reactions leading to synthesis of organic compounds, specially carbohydrates from the O_2 of water and carbon of CO_2 in presence of light as the energy source in chlorophyll containing tissues of plants.

Physiological loss: Lowering down the quantity of the perishables due to normal physiological process after harvest and *e.g.* toughening and spongyness in green beans, sweet corn, carrot, radish etc.

Physiological loss in weight: Weight loss of fresh fruits and vegetables attributed to many physiological activities like transpiration, respiration etc. Which depends upon the temperature, relative humidity, air circulation and other storage environment.

Physiological maturity: Attainment of final stage of biological function by plant part of the plant as a whole.

Phytolexin: A phenolic substances having antifungal principle synthesized by plants in response to parasite invasion or infection by certain fungi *e.g.* pisatin, phaseolin, trifolirhizin, orchinol and isocumarin from pea and bean pods, red clover, tubers and carrot, root respectively.

Pie diagram: A diagram having the shape of circle whose radius is proportional to the square root of the aggregate of values.

Piezometer: Instrument used for measuring depth of water tables.

Pinching: Removal of the terminal growing portion which reduces the height but promotes axillary branching, delays flowering and helps in breaking rosetting.

Pioneer species: Plant forms first developing on a site after complete destruction of the previous flora.

Plant container: Containers in which plants are raised from seed or into which they are transferred from seed bed for the purpose of planting out later.

Plant growth regulator: Any hormone or synthetic organic compounds capable of causing physiological response in plants.

Planting box: A box or tray in which small plants are raised and transported to the field for planting out.

Planting ratio: The male and female plants when planted in a certain proportion to ensure proper pollination and fertilization *e.g.* 10 : 1 (Female : Male) ratio in pointed gourd.

Plant lifting: Removal of plants from nursery beds with or without earthball for sale, replanting etc.

Plant propagation: Perpetuation and multiplication of plants.

Plumule: Primary stem-bud of an embryo as it develops from the seed in germination.

Pod: Dry, one chambered fruit developing from a simple pistil and dehiscing by both the margins *e.g.* pea, bean etc.

Pollen: The male reproductive unit produced within anther labies.

Pollen compatibility: Ability of the pollen tube to grow through a style and reach the ovule in time to effect fertilization.

Pollen tube: The structure produced by the germinated pollen grain which grows down the style and normally into the embryo sac through the micropyle and through which male nuclei is carried to the ovule for fertilization.

Pollination: The transfer of pollen from an anther to a stigma.

Pollinators: The agency by which pollen is transferred from an anther to a stigma.

Polyembryony: Occurrence of two or more embryos within a single seed which may result from nucellar embryony, occasional development of more than one nucleus within the embryosac and cleavage of the proembyos during the early stage of development.

Polyphenol oxidase: An enzyme causing oxidation of phenols or polyphenols which may be the principal reaction in enzymatic browning. Removal of oxygen by blanching or sulphiting of material or deaeration of juice can help to prevent the activity of this enzyme.

Population: All the possible individuals or observations of one kind from which samples are drawn for statistical analysis.

Porogamy: Penetration of the pollen tube through the micropyle.

Post harvest physiology: Physiology of vegetables after harvest because they behave like living organisms and continue to perform the metabolic reactions (respiration, transpiration) and maintain the physiological system which were present when they were attached to the plant and for this metabolic processes, they depend entirely on their own food reserves and moisture content.

Post harvest technology: The techniques adopted to increase the shelf-life of harvested produce which includes packaging, use of low and high temperatures, chemical and other suitable methods.

Pot herbs: A group of vegetable crops whose foliage and sometimes immature stem are used as cooked vegetables also called leafy vegetables or green *e.g.* palak, amaranthus, spinach, basella etc.

Potting compost: Compost specially prepared to make a suitable medium for the plants to be planted in pots.

Precision: Relative or apparent nearness to the truth.

Pre-cooling: An important post-harvest operation of vegetables which is accomplished by refrigeration or water immersion with a view to removing field heat, conserve weight and retard ripening and senescence of the harvested vegetables.

Pregermination: Germination of seeds, usually until the radicle just emerges/before sowing.

Pre-processed spoilage: Spoilage of products (raw materials) in cans before processing which is caused by all types of micro-organisms.

Preservation: Methods of extending the shelf-life of fresh as well as processed products by means of low temperature, drying, anaerobic condition, chemicals, irradiation, fermentation etc.

Pricking off: Removal of seedlings from beds.

Primary data: The data which the investigator originates for the purpose of a specific enquiry.

Primary dormancy: The conditions that exist within the seed to prevent germination at the time it matures on the plant and in the immediate period afterward.

Primary leaf: Just true leaf which are unifoliate and opposite in some legume crops like cowpea.

Productive stage: The period of fruit fullness, productivity or accumulation of carbohydrates.

Productivity: Capacity of a crop to produce yield or biomass per unit area per unit time.

Productivity index: It gives an overview of economic yield obtained from any crop *e.g.*

$$\text{Productivity Index of Pea} = \frac{\text{Av. wt. of green grains/plant}}{\text{Av. wt. of green pods/plant}} \times 100$$

Psychrometer: Instrument used to measure the humidity of the atmosphere.

Pubsescence: Short and soft hair covering on the plant parts.

Pulper: A machine that chops up whole fruits or vegetables and screens out the large particles and seed to form a pulp that can be used directly of fed into the finisher to obtain juice.

Pure crop: A crop composed almost entirely of one kind usually to the extent of not les than 80 per cent.

Pure seed: The seed which is true, to its kind or variety.

Quantitative character: Characters showing continuous variation and nature of inheritance of this character is conditioned by polygenes each having small and similar effect. These are also called metric characters.

Quercetin: A pigment which imparts coloration in the outer skin on onion bulb.

Quiescence: The condition in which the seed can germinate immediately upon the absorption of water in the absence of any internal germination barriers.

Qualitative character: Character showing distinct classes and nature of inheritance of this character is governed by one or few major genes each having large and distinct effect.

Quick freezing: Freezing of products in a relatively short time (30 minutes or less) may be done by direct immersion of the food (packed or unpacked) in a refrigerant (liquid nitrogen or liquid CO_2) or by indirect contact with the refrigerant (air blast freezing).

Quadruple cropping: Growing four crops a year in sequence.

Quadrat: A small field study unit or sample area, usually a square metre or a milliacre size, established for the purpose of detailed observation.

Raceme: Inflorescence where the main axis is elongated and bears laterally a number of flowers which are all stalked, the lower flowers having longer stalks than the upper ones, as in radish, caesalpinia etc.

Racemose inflorescence: Syn. Monopodial inflorescence.

Rachis: The flower-bearing portion of an inflorescence.

Radicle: The lower portion of the embryo axis which give rise the roots.

Randomized block design: An experimental design where the number of blocks are equal to the number of replication of each treatment and each block is divided into homogenous plots of equal size within which the treatments are applied randomly.

Randomization: Allocation of treatments to various experimental units in such a way that each treatment has an equal chance of being allocated to the experiment unit.

Range: The simplest measure of dispersion and is defined as the difference between the largest and the smallest item of the series.

Real value of seed: Percentage of a seed sample that would produce seedlings of the variety under certification and can be determined as Purity per cent × Germination per cent 100.

Recalcitrant seeds: A category of seeds which require relatively high moisture content for longevity and when dried below a certain critical moisture level, they rapidly lose viability and these seeds are usually short lived *e.g.* Seeds of many fruit, plantation and timber crops.

Reclamation: The operation of changing the condition of soil which cannot be utilized to its full potential otherwise *viz.* gypsum application to alkali soils, liming of acid soils etc.

Refractometer: An instrument used for measuring the refractive, index of lipids. Total soluble solids are estimated by their instruments.

Refrigerated storage: Storage at temperatures below those normally available throughout the year, generally between about 0°C to 8°C.

Regeneration (Tissue culture): One of the major manifestations of the juvenile to mature phase changes where new organs (roots, shoots or embryos) regenerate an explants and propagules in tissue culture system.

Registered seed: It is the progeny of foundation seed that is so handled as to maintain satisfactory genetic identity and purity and that has been approved and certified by a certifying agency. This class of seed is used for the production of certified seed.

Regression analysis: A statistical analysis for associationship which aims at estimating or predicting the unknown values of dependent variables for known values of independent variable.

Relative growth rate: A measure of the productive efficiency of a plant which is expressed as the grams of dry weight increase per gram of dry weight present per day.

Relative humidity: The ratio of the weight of water vapour in a given quantity of air to the total weight of water vapour that quantity of air is capable of holding at a given temperature, expressed as percentage.

Relative yield index: Mean crop yield of a region divided by the mean all India yield of the some crop times hundred.

Replication: Repetition of treatments in experiment to increase the precision of comparisons.

Ring methods (irrigation): A modified basin method of irrigation in which flat basins around the tree trunk are connected individually with a common channel passing between the rows of plants and water is let in through the common channel to the basin only.

Roguing: Removal of off-type or undesirable plants from the population to maintain purity of the crop during seed production.

Root cap: Each root is covered at the apex by a short of cap or thimble known as the root cap which protected the tender apex of the root as it makes its way through the soil.

Root crops: A group of vegetable crops whose swollen tap roots and in some cases, hypocotyl along with the tap root such as carrot, beet, radish, turnip, rutabaga etc. are cooked or eaten raw.

Root forking: Branching of tap roots in the root crops particularly in radish and carrot due to the presence of impediment, undecomposed organic matter or plant refuse in the soil.

Root to seed method: A method of seed production in root crops where the fully matured roots are harvested, selected and after giving proper root and shoot cuts, they are replanted for seed production.

Run off: Portion of precipitation which form the surface flow.

Saline soil: The soil is said to be the saline if it contains an excess amount of soluble salts leading to an increase in osmotic pressure of the soil which retards plant growth.

Salinization: The process of accumulation of salts in the soil.

Salometer: An instrument used for measuring the salt concentration of any substances is terms of degree salometer.

Salt respiration: The increase in respiration upon transfer of a tissue or plant from water to salt solution.

Sample: A part of the population selected as a representative to the whole.

Sample plot: A plot chosen as representative of a larger area.

Sampling: The process of selecting at random a small part of anything intended as representative of the whole.

Sampling error: Deviation of a sample value from that of the population due to small size of the sample.

Sandy loam soil: The soil which contains 50-80 percent sand and 20-50 percent silt and clay with 0.1-1.0 percent organic matter.

Saponine: A biologically active substances present in a number of vegetable crops like spinach, tomato etc that promotes intensive growth and development of the accompanying plants.

Scarification: Any process of breaking, scratching mechanically altering or softening the seed coverings to make them permeable to water and gases.

Scooping: Removal of central portion of the surd for easier initiation of flower stalk in cauliflower.

Seasonal consumptive use: The total amount of water consumed in evaporation and transpiration by a crop during the entire growing season expressed in depth or volume of water per hectare.

Secondary dormancy: Dormancy that develops within the moist seed after it is removed from the plant and is subjected to adverse environmental conditions.

Seed: A matured ovule that consists of an embryo, its stored food and protective convering it also commonly includes the ovules of one seeded dry indehiscent fruits like paddy, wheat etc.

Seed bed: A prepared area in which seeds are sown.

Seed certification: A legally sanctioned scientific system for quality control of seed multiplication and production which consists of the following control measures: administrative check on the origin of propagating material, field inspection, sample inspection, bulk inspection and control plot testing.

Seed coat dormancy: Dormancy of seed that is controlled by the seed coat.

Seed drill: An implement for sowing seeds in lines.

Seedling: The juvenile phase of a plant grown from seed.

Tap root: Root that is developed from the radical directly giving rise to the main root of the plant.

TA spoilage: Spoilage of low and medium acid canned foods by *Clostridium thermosaccharolyticum* which breaks sugars and forms acid and gas (Mixture of H_2 and CO_2) resulted in swelling and in severe cases bursting of can may occur and the spoiled food usually has a sour odour.

Taxonomy: The science of identification, nomenclature and classification of organisms.

Tegmen: Inner layer of seed coat.

Tenderometer: An instrument by which toughness of the seed coat and firmness of pulp is determined and is mostly used to determine the seed quality of pea where high value of tenderometer indicates low quality.

Tendril: Slender, filiform, spiral or hooked organ arising from stem or foliage, affording plants a means of obtaining support by climbing trees or other bodies.

Tensiometer: A device for measuring the negative pressure (Tension) of water in soil *in situ*. It is a porous, permeable ceramic cup connected through a tube to a manometer or vaccum gauge.

Testa: Outer layer of seed coat.

Tetrazolium (TZ) test: a biochemical method for determining the viability of both dormant and non-dormant seeds by the appearance of red colour when the seeds are soaked in 2, 3,5-triphenyltetrazolium chloride solution.

Thermodormancy: A type of physiological seed dormancy where germination of the freshly harvested seeds is sensitive to temperature *e.g. Streptococcus, Lactobacillus* etc.

Thinning: Removal of extra seedlings flowers or fruits to avoid injury due to overcrowding.

Thinning out: Type of top pruning where the entire twig, cane or shoot is removed.

Tillage: Working the soil to provide condition suitable for plant growth.

Tilth: The physical condition of the soil in its relation to plant growth.

Topo-climatology: Study of local climate affected by slope and other surface characteristics.

Topography: Configuration of surface including its relief and the position of its natural and man made features.

Top soil: The surface soil, usually up to the plough depth (15-20 cm from the surface).

Town compost: Compost which is made from town refuse like night soil, street sweeping and dustbin refuse.

Transplant: A seedling or rooted cutting after it has been moved one or more times in a nursery in contrast to a seedling planted out directly form the seed bed.

Transplanting: Removal of seedling or plants from one place and planted in other place.

Trickle irrigation: Syn. Drip irrigation.

Trifoliate: A compound leaf with three separate leaflets as in French bean, cowpea etc.

Truck gardening: Specialized type of vegetable farming where one or two special vegetable crops are grown usually in large quantities suited for the particular region.

Undesirable plants: Plants which confirm to the varietal characteristics but not acceptable due to undesirable features.

Utilization index: Ability of root and tuber crops for better accumulation of photosynthates in the storage organ which can be judged by root : shoot ratio.

Vaccum drier: A drier used to dry heat sensitive products where low pressure lowers the temperature which helps to retain natural flavor and minimizes oxidation and browning.

VAM: It refers to vesicular arbuscular mycorrhiza which play important role in the growth of many crops like pea, cowpea etc.

Vegetable: An edible plant or plant part eaten cooked or raw as a main part of a meal, side dish or appetizer.

Vegetable forcing: A specialized type of vegetable farming where vegetables are grown out of their normal season. Vegetable forcing requires some special structures like glass house, hot beds, cold frame etc.

Vermiculite: Micaceous mineral and chemically hydrated magnesium aluminium silicate very light in weight, insoluble in water and is able to absorb large quantities of water.

Vertical mulching: Incorporation of vegetative mulch in a band for the purpose of harvesting rain water and conserving soil and water.

Viability: The ability of pollen seed etc to live grow and develop.

Viable seeds: Seeds capable of producing normal healthy seedling under favourable conditions.

Vigour index: A measure of seedling vigour calculated by the formula: germination per cent x mean length of root + shoot of the seedling and expressed as a whole number.

Viscosity: Internal forces of friction between layers of fluid by virtue of which a fluid resists its flow over a surface, inherent property of all fluids.

Volume percentage: The number of parts of aliquid substances per 100 parts of the substance by volume.

Volunteer plants: Unwanted plants growing from seeds that remain in the field from a previous crop.

Waste land: Land not capable of producing crops or of giving services of values.

Water availability: It may be expressed in terms of available water for growth of micro-organisms (depends on kind of micro organisms) which is the ratio of vapour pressure of the solution (of solutes in water in most food) to the vapour pressure of solvent (usually water). For pure water aw is equal to 1.00.

Water harvesting: Conservation of rain water under unirrigated condition by collecting runoff of precipitation in order to supplement soil moisture in an adjacent area.

Water holding capacity: The weight of water held by a given quantity of absolutely dry soil hen saturated.

Water logging: A condition in consequence of inadequate drainage where the soil pores get filled with water by the exclusion of air.

Water table: The upper surface of the zone of saturation of soil by ground water.

Weed: Any plant in a place where it is a nuiscene.

Whippy seedling: A seedling which has not yet developed a rigid woody stem.

Wound roots: Adventitious roots which are developed only after the response of the wounding effect in preparing the cutting.

Xanthophyll: Yellow, hydroxyl carotene derivatives ($C_{40}H_{55}O_2$) which is present in all green leaves together with chlorophyll and carotene.

Xerophytes: Plants which are adopted to live in drought conditions.

Yearling: One year old bulblets which has been formed on the scale.

Yield component: The components or attributes which finally make up or control yield of any crop.

Zero selection: Selection with highest intensity when the proportion of the population rejected is 100 per cent.

Zero tillage: Crop growing without tillage which is involved in higher water content in the top soil layer, stronger mechanical resistance to root penetration, less soil temperature amplitude and different pattern of nutrient distribution in soil profile.

Zoning of beet: Under unfavourable conditions particularly in hot weather, beet root show alternate white and coloured circles when sliced, called zoning.

Suggested Reading

Bose,T.K. and M.G. Som (1986). Vegetable Crops in India, Naya Prakash. 206, Bidhan Sarani, Kolkata 700006.

Biswas and Biswas (1990). Kitchen Garden, Chash-Bas, Konnagar, Hooghly, W.B.

Choudhary, B. (1990). Vgeetables, National Book Trust, India, A-5 Green Park, New Delhi-110016

Chauhan, D.V.S. (1968). Vegetable Production in India, Ramprasad and Sons, Agra-3

Das, P.C. (2008). Vegetable Crops of India. Kalyani Publishers.

I.C.A.R. Hand Book of Horticulture.

I.C.A.R. websites.

IIVR Annual Report.

Salaria nd Salaria (2007). Horticulture at a Glance, Volume II. Jain Brothers, New Delhi.

Thomson, H.C. and W.C. Kelly (1957). Vegetable Crops. Tata McGraw Hill Publishing Co. Ltd., New Delhi-110002.

Yawalkar, K.S. (1969). Vegetable Crops of India. Agri. Horticultural Publishing House, Agra.